BORNE ON THE WIND

To Ron,
who many times put
the camel through the
eye of the needle

BORNE ON THE WIND

The Extraordinary World of Insects in Flight

by

STEPHEN DALTON

Edited by John Kings

Chatto & Windus   London   1975

ACKNOWLEDGMENTS

I would like to express my thanks to the following, who helped me in various stages in the preparation of this book:

To Kodak Ltd, London; Kodak, Inc., Rochester, New York; Leitz Instruments Ltd, Great Britain; to Michael Tweedie for his entomological advice; to my sister, Tessa, who helped me greatly in Everglades National Park; to the U.S. National Park Service and to the Superintendent of Everglades National Park; to Professor Howard Ensign Evans of Colorado State University, who was kind enough to write the Foreword; to Professor Torkel Weis-Fogh of Cambridge University for allowing me access to his latest research on the aerodynamics of insect flight; and, finally, to my editor, John Kings of *Reader's Digest Press*, New York, for his continuing enthusiasm and creative interpretation of the subject.

Published by
Chatto & Windus Ltd
40 William IV Street
London WC2N 4DF

ISBN 0 7011 2130 0

Production Supervision by VISUAL BOOKS, Inc.
Designed by Nicolas Ducrot

Color separations by Offset Separations Corp.
Typography by Typographic Art Incorporated
Printed by Eastern Press, Inc.

CONTENTS

LIST OF PLATES

ODONATA: DRAGONFLIES
AND DAMSELFLIES

ORTHOPTERA: GRASSHOPPERS
AND THEIR RELATIVES

HEMIPTERA: BUGS

NEUROPTERA AND MECOPTERA:
LACEWINGS AND SCORPION FLIES

LEPIDOPTERA: BUTTERFLIES AND MOTHS

DIPTERA: TRUE FLIES

FOREWORD

The flight of insects has intrigued curious persons since the time of Leonardo da Vinci, and doubtless before. Leonardo and, almost two centuries later, G. A. Borelli propounded a number of aerodynamic principles, though they based most of their conclusions on a study of birds. Even Louis Agassiz, after the passage of another two centuries, incorrectly supposed that the wings of butterflies and moths became separated during the upstroke, allowing air to pass between them as it does between the primaries of the wings of birds.

With increasing knowledge of insect anatomy, and particularly with the invention of high-speed motion-picture photography, the unique properties of insect flight gradually became more apparent, for insects are too small and their wingbeat too fast to permit many conclusions from direct observations. Still, there remained many controversies. Puzzlement over the mechanics of insect flight led certain researchers to conclude that some new principles must be involved, and this led to the often repeated (but not quite correct) statement that biologists have proved that insects really can't fly! Then there was the episode concerning an American entomologist who, as recently as 1926, watched deer botflies flying past him with such speed that they were "only a brownish blur . . . and without a sense of form." He estimated their speed at 820 miles an hour (more than the speed of sound), a report so remarkable that it reached the pages of *The New York Times* and the *Illustrated London News*. Twelve years later the physicist Irving Langmuir calculated that propelling a fly at this speed would consume fuel so rapidly that the whole insect would burn up in half a second. And if a botfly hit a person at that speed it would penetrate with the force of a bullet! In fact, we now know that botflies, although among the fastest of insects, achieve a speed of no more than about twenty-five miles per hour.

Fortunately we are now well past the stage when anyone is likely to make so gross a miscalculation, but insect flight is still a very active field of research, as Stephen Dalton's book makes clear. We know, for example, that the exceedingly rapid contractions of the flight muscles generate considerable heat, so that the temperature inside the thorax is sometimes as much as 20 degrees Celsius above the outside temperature. In fact, heavy-bodied insects such as bumblebees and hawk moths require flight muscle temperatures of 30 to 40 degrees C before sufficient lift can be produced to support them in flight. This raises some serious questions: How do they prevent overheating? And how do they manage to take off when the air temperature is below 30 to 40 degrees C (86 to 104 degrees Fahrenheit)?

Bernd Heinrich of the University of California at Berkeley recently showed that when the dorsal blood vessel of a hawk moth is tied off just behind the thorax, the thorax does, in fact, overheat, so that the moth cannot be induced to continue flying. Evidently the blood passing back from the thorax is cooled in the abdomen before being returned to the flight muscles via the dorsal vessel. And it has been known for some time that heavy-bodied insects such as hawk moths and giant silk moths must warm up their flight muscles by "shivering" before they are able to take flight. It is noteworthy that those insects in which temperature regulation is critically important have bodies densely clothed with scales and hairs, providing a measure of insulation.

The flight of the bumblebee is a particularly remarkable performance. Even when temperatures are barely above freezing, they are able to achieve and maintain an internal temperature of more than 30 degrees C. Bumblebees weigh only one- or two-tenths of a gram, and for an organism so small to produce so much heat requires a prodigious expenditure of energy, which is fueled by nectar from flowers or honey from the pots in their nests. Bernd Heinrich and his co-workers have shown that the bumblebee is somehow able to uncouple its wings from its flight muscles when it is stationary, so that it can vibrate and thus warm up without actually moving the wings. Bumble-bees, like some other bees and wasps, also use body heat so generated to raise the temperature of the larvae and pupae in their nests.

Although bumblebees do not have a great many predators, there is no doubt that many moths that require a prolonged warm-up before flight are especially vulnerable to attacks by birds. It is noteworthy that many of these moths, such as hawk moths and giant silk moths, have large eyespots on their wings. It has been shown experimentally that these eyespots do, in fact, have survival value in frightening off birds, especially when combined with the shivering movements that are prerequisite to flight. The British behaviorist A. D. Blest showed that artificial eyespots, even when painted on quite abnormal subjects, such as mealworms, greatly reduced the rate of predation by several kinds of birds.

All of this serves to remind us that insects use their wings for many purposes other than flight. Some have wing shapes and colors that blend with the background, rendering them virtually invisible. Many butterflies have spots along the edge of the wing which serve as deflection marks, inducing birds to nip them there rather than in more vulnerable parts of the body. Some of the hairstreak butterflies even have a "false head" formed from the back part of the hindwing, presumably causing birds no end of confusion.

We tend to think of the scales of a moth's wing as serving primarily to produce the color pattern. But Thomas Eisner of Cornell University has shown that the relatively loose scales of the wings of small moths enable them to escape from spiders' webs and from insectivorous plants such as sundews. When captured, they merely struggle a bit and work their way out, leaving a few scales behind. The looseness of the scales of many moths is shown beautifully in several of Stephen Dalton's photographs, in which moths are seen taking flight and leaving a cloud of scales behind them.

Honeybees and some other social insects also engage in fanning at the nest entrance during hot weather, thus providing a measure of air conditioning. Stephen Dalton mentions a small wasp that parasitizes the eggs of aquatic insects and uses its wings to swim through the water.

Above all, one thinks of the many ways that wings are used to provide signals that assist in bringing the sexes together—and reproduction is, above all, what insects are best at. Even as I write this, in November, short-horned grasshoppers are clattering in my back yard, the wing sounds of males still hopeful despite the imminence of snow. In male crickets and katydids the front wings are greatly modified for producing some of the most remarkable sounds in the animal kingdom.

Even the simple wing sounds of midges and mosquitoes play a role in bringing the sexes together. In this case it is the female that attracts the male by the hum of her wings, a fact quickly apparent to singers who hit a G in the vicinity of a swarm and end up with a mouthful of male mosquitoes. Certain male fruit flies use their wings to fan toward the female a chemical sex attractant produced by their glands. Many male butterflies have parts of their wings covered with specialized scales that produce substances attractive to the females. When these are removed, the males court even more vigorously, but with frustrating results.

Of course, wing colors may also play a role in courtship, but less often than you might think, for most insects are better smellers than they are seers. One exception is provided by the dragonflies and damselflies, which have huge eyes and very small antennae. Here we know that colors and the translucence of flickering wings do play a role in courtship. In butterflies there are subtleties we have only just begun to appreciate. Robert Silberglied of Harvard University, along with several co-workers, has found that the males of certain yellow "sulfur" butterflies have patches of scales that reflect ultraviolet light. These patterns are invisible to us under normal conditions, but insects perceive ultraviolet, and it is clear that these patterns play a role in bringing together the two sexes of one species.

The wings of certain insects even play a role in the mating act itself. In at least one species of cricket, the female consumes the wings of the male during copulation. Perhaps the most remarkable instance of the use of the wings during mating involves a curious species of scorpion fly that occurs on patches of moss during the colder times of the year. These insects are flightless, the female having no wings at all and the male having wings that are somewhat rod-like, with a hook at the end. The male uses these hooks to seize the female and later to hold her in the proper position for copulation.

But of course by far the most important function of wings is to fly, and that is what this book is all about. As Stephen Dalton points out, insect wings are unique in at least two major ways. For one, there are no muscles in the wings as such, and all the power and control emanate from the body. For another, wings are not modified limbs as they are in birds and bats, and as they were in pterodactyls, but entirely new structures arising from the back —a feature, as some wag has pointed out, that insects share only with angels.

Insects had been airborne for many millions of years before birds and bats appeared and began to exploit them. These new aerial predators presented not so much a threat as a challenge to insects, resulting in the evolution of swifter and more efficient flight patterns, as well as special mechanisms for avoiding capture. Bats, it is well known, emit ultrasonic chirps which bounce off objects in their path and inform them of the presence of obstacles as well as potential prey. Several groups of moths have responded by developing receptors for ultrasound. Kenneth Roeder and his associates at Tufts University have shown that these moths not only detect the bats' sonar but respond by abruptly changing their course of flight, thus frequently avoiding capture. Tiger moths have carried this one step further, having themselves developed the ability to produce ultrasonic clicks. These are produced in response to those of the bat, and evidently tell the bat "I am not good to eat," for in fact these moths have also developed nauseous fluids that render them distasteful.

Certain species of flies not only resemble yellow jackets closely in shape and color but even exhibit audio mimicry, producing an identical hum when they fly because the rate of wingbeat is identical. It has been shown experimentally that birds avoid yellow jackets because of their sting and thereafter avoid other insects resembling them.

How did insects first evolve those unique structures, the wings, and evolve them so early in geologic time? Here we enter another area of controversy, and one that is not likely to be resolved right away. When I was a student I learned that there were two theories: the flying-squirrel theory, that wings arose as flat outgrowths first used for gliding; and the flying-fish theory, that wings were originally plate-like gills that were used for flopping about on land when the primeval fresh water pools dried up. The latter view is suggested by a look at the immature stages of mayflies, which today have plate-like gills on the abdomen that are moved by muscles. But the flying-squirrel theory has been much more popular down through the years and is supported by the fact that some of the ancient fossil insects have broad flaps on the back of the thorax that look like gliding surfaces and incipient wings. Stephen Dalton discusses this theory in more detail in his chapter "The Flight of Insects."

Just recently the British physiologist Sir Vincent Wigglesworth has revived the flying-fish theory in slightly different form. Wigglesworth points out that one group of terrestrial insects, the very ancient bristletails, although wingless, have small "styli" arising from the thoracic leg bases and the underside of the abdomen, and these are homologous to the plate-like gills of mayflies. He points out that recent research has shown that the air is filled, during the summer, with myriad insects, the "aerial plankton," often high in the sky and far out to sea. Most of these are winged, but wingless forms, even bristletails, have been taken high in the air. During a period of aridity, back in the Paleozoic era, the originally aquatic bristletail-like insects might well have joined the aerial plankton and learned to fly, not, he points out, "by studying aerodynamics, but by trial and error," the thoracic styli providing the materials from which wings evolved.

Still another theory has been proposed by two American entomologists, Richard Alexander of the University of Michigan and William Brown of Cornell. They believe that wings may have arisen in a terrestrial environment and as accessories to mating. Perhaps they were originally flaps to cover certain glands on the back that produced a sex attractant (as in tree crickets today); they may then have become enlarged and brilliantly colored, providing an advertisement in courtship and territorial defense, analogous to the dewlap of many lizards. Since this theory was proposed, it has been learned that some of the very earliest insects preserved as fossils did, indeed, have vivid color patterns on their wings.

Wherever the truth may lie, it is obvious that the acquisition of wings unlocked a whole new world for the insects, which might otherwise have remained as insignificant a part of the environment as, for example, their relatives the centipedes. We believe that there may be a million species or more, and, according to one estimate, the total insect population of the earth at any one time is a billion billion. Insects occur from Antarctica to well north of the Arctic Circle; from deep in the soil and the bottoms of deep lakes to the tops of high mountains and more than two miles high in the air. Even the most remote oceanic islands have a rich insect fauna.

It was once believed that much of the dispersion of insects was the result of accidental drifting about in air currents. We now know that there are episodes in the life cycle of many insects, often soon after they acquire their wings and adult reproductive organs, when they deliberately launch themselves into the air and undertake flight patterns likely to take them to appropriate places to feed and reproduce. When there are mass flights of large insects, such as locusts, we can well appreciate them, but the migrations of plant lice, thrips, and myriad other small insects go largely unnoticed except by the specialist. The migration and dispersal of insects by flight is such a large and rapidly developing field of research that it took C. G. Johnson of the Rothamsted Experimental Station in England a book of 763 pages to survey the field, and doubtless he would be the first to admit that much more could be said on the subject.

The distances that insects are able to fly border on the incredible. Mosquitoes and blackflies may occasionally range as much as fifty to one hundred miles from their breeding places, when aided by the wind and when they are able to refuel by feeding on nectar along the way. Some insects are capable of long, sustained flights without refueling; flights of dragonflies have, for example, been sighted many hundreds of miles at sea. The most remarkable, well-documented long-distance flight of an insect concerns the small mottled willow moth, which invaded England in great numbers in May 1962. Using data gathered by his staff and by amateur entomologists, combined with a study of weather maps for the period just preceding the invasion, C. B. Williams determined that the moths had flown nonstop over the sea from Morocco, a distance of about two thousand miles, in four days.

In spite of all this, it must be admitted that the average person is barely aware that insects exist at all, except perhaps for the fly that trespasses on his patio or the aphids that speckle the house plants. Stephen Dalton has developed a method of taking color photographs of insects in free flight, untethered and without artificial stimulation—the first successful undertaking of its kind—that will rank as a most exciting visual aid to entomological studies. He explains his techniques in his final chapter, "The Photographer at Work." Dalton is about to convince you that insects do indeed exist, that as flying machines they are unique as a group, yet diverse almost beyond belief, and above all, that they are things of unsurpassed beauty. I, for one, have always been willing to put up with occasional mosquito bites and wormy apples, such is my respect for the creatures that cause them, and such is my hope for a world of coexistence not only among men but among all living things.

Howard Ensign Evans
Fort Collins, Colorado
November 1974

PREFACE

At the age of about seven I lost my heart, and often, I think, my reason, to three interests that were to become the obsessions of my life. These separate passions—photography, flying, and insects—ultimately coalesced into a preoccupying, shimmering challenge that has dominated the past ten or twelve years of my otherwise relatively normal existence.

My introduction to photography was dramatic. One day my father sought to satisfy my inquisitive, ferreting nature by confining me to an elaborate blind he had built from which to photograph one of his favorite birds. "Stay there, keep quiet, keep looking, and when the kingfisher appears, press this button," he instructed.

Crouched in that secret place, confronted by an antediluvian plate camera on a mammoth tripod, I achingly awaited my quarry. And when the resplendent kingfisher swooped into view, I dutifully and joyously pressed the release. The resulting blurred image of that lovely bird was enough to make me an avid photographer of nature from that day on. Subsequently I was rewarded by a gift of my first camera, a box Brownie from which I immediately expected the most advanced results.

My interest in flying is a legacy from World War II. My father was in the R.A.F., so that I was always conscious of planes and wings and blue uniforms, and of bombers droning through the night sky of the English countryside.

We were living then in the Midlands, near an airfield, and I would go close to its high wire fence and watch for hour after hour the planes take off and land. I was not really concerned, I think, with where they were going or where they had been, or with their missions of destruction, but rather with the precision and power and style of their flying. Soon I knew that I also wanted to be a pilot.

Later, much later, I qualified for a private pilot's license, experiencing in my own flying a fraction of the aerial sensations that only nature's creatures can know in all their spectacular diversity.

The third interest was the most predictable of the three. My whole family—

my father, mother, sister, and brother—lived with an intense awareness of nature and living things. Dogs, cats, horses, birds, animals of the field and forest, anything that approached for affection, ran off in fright, crawled, flew, or froze in its tracks became the object of our care and affection.

We all shared these interests, but my own particular focus gradually narrowed to insects, so that, relegated to the long grass of the outfield at school cricket matches, where my ineptitude at catching the fleeting ball was least likely to become apparent, I would spend my time watching a fly on a blade of grass, or a swarm of gnats hovering over the bordering hedge. Acidly, my school athletic report one year commented: "Dalton is more interested in the lilies of the field than in the pursuit of the game."

My godfather encouraged my curiosity concerning insects by showing me his collection of butterflies, moths, and beetles, and the highlight of school vacations was to visit him and pore over the cases of his prized specimens, marveling at the translucency of wing and the perfection of design that enable these tiny creatures to fly with such superb technique.

My early efforts at photography were ludicrous. Wanting to capture the image of a greenfly on a rosebud, I pushed the lens of my box camera within an inch of my prey, expecting to achieve in this manner a large, sharp picture. Happily I romped from bush to flower, all around the garden, shooting every insect I could find. The resulting totally blurred impressions made my first lesson in the limits of photographic equipment shattering and indelible.

Over the years my knowledge of insects grew, and my photography improved to the point where I could capture their quality at rest on a leaf. But despite various experiments with high-speed flash, the most intriguing aspect of insect life continued to elude my camera, as it did my eye. There was no way in which I could observe the exact way in which an insect flies, how it uses its wings to make the incredible maneuvers we take for granted.

There simply did not exist any photographic technique capable of stopping with absolute clarity the wingbeat of an insect in free flight.

It was the solution to this problem that, in my professional career as a photographer, became an overriding obsession. My three childhood interests had fused into one tantalizing quest, which has now become the subject of this book.

S. N. D.
Surrey, England, 1975

THE FLIGHT OF INSECTS

Despite a generally held suspicion that Nature placed insects on this earth specifically to chastize the delicate and sensitive hides of men, entomologists might well claim the reverse, for they have established that insects were prior tenants by some 350 million years. Moreover, since insects are an essential link in the chain of life, without their contribution our own existence would not be possible.

That our success as a species depends upon the whim of a fruit fly is a hard debt to acknowledge—for it carries no overtones of heroic mission—yet it is undeniable fact. To put it bluntly, the world is a meadow made for insects, not for men. It is a meadow used by at least a million separate species of insects for take-off and landing in a nonstop aerial display of flight techniques and aerobatics so stunning and spectacular that, in comparison, even the most advanced manned flight is but the first faltering step of a child.

Man's only original contributions to developments in air travel are confined to ballooning and rocketry, neither of them relying solely on true aerodynamic principles for their lift, and neither very clever when contrasted with the prowess of a gnat. But man's efforts at change are at least conscious, thinking endeavors, and that of course is what sets his development apart from the evolutionary pattern of insects.

In a way, man precipitates his own evolution, designs it materially, makes it out of string and sealing wax, is not content to wait fifty million years to grow webbed feet or a spread of wings. He demands instant change, instant results, instant "progress." And if that is his intended role, to use his intelligence creatively, how noble it is and yet how sad that in the process avarice so often blinds. Perhaps man is too clever by far, juggling the environment, the forces of nature, and his own condition in a precarious act that could not only be the greatest but possibly the last show on earth. Perhaps there is much to be said for the infinitely slower development that Nature prescribed for our supposed inferiors.

Insects were the first creatures to fly, long before birds or bats, though fossilized evidence of those early species is more rare than of vertebrate animals, due to their fragility when compressed between the layers of sediment and time. It is generally agreed, however, that the wings of insects probably developed as extensions of the thorax wall, first as shallow extrusions that helped stabilize them as they jumped out of harm's way, later becoming wider, rigid, flat surfaces that helped them glide to the ground. Much later, insects developed the muscles in the body wall that enabled them to flap these flat surfaces, turning them into wings.

It may well have taken more than 100 million years for them to become truly airborne, and perhaps one day the links will be exposed in coalbeds churned up to satisfy our needs. Meanwhile it is enough to know that the wing evolved, that original insects did not come ready-made for flight, and that they show a continuing capacity for change, sometimes detectable even in the measurable span of an entomologist's own lifetime.

For example, even for an aphid to build resistance to an insecticide spray is essentially an evolutionary refinement, and the much-quoted instance of the peppered moth in the smoky English Midland towns mutating its color pattern from pale gray to a dark sooty gray in order to preserve its camouflage is convincing proof that Nature looks after its own, usually in all the slowness of extended time, but sometimes, when necessary, very fast indeed.

The flying ability of insects has been the main factor in their success as a class, enabling them to disperse farther afield to colonize new areas, escape danger more easily, and look for food when locally it becomes scarce. Flight has made insects most efficiently mobile, in a menacing world where mobility has always been the name of the game, whether for man or moth.

The early flying insects were somewhat rudimentary in pattern. A good example of antique flight structure is the dragonfly, virtually unchanged from its Pennsylvanian ancestor of 350 million years ago, but certainly successful; otherwise it would not have survived. In the Carboniferous period, some 300 million years back, primitive cockroaches, mayflies, dragonflies of all sizes, and clumsy Palaeodictyoptera—the "ancient net wing"—which did not survive as an order and should probably never have been allowed off the drawing board, had the primal air lanes to themselves. Swooping and dancing along the edges of lagoons and lapping seas, these early fliers logged more than 200 million years of flying time before the first test flights of the birds, making them airworthy by any standards.

The wings of insects comprise two thin layers of chitin—a thin shell-like substance—which are sandwiched together like two sheets of plastic wrap and strengthened by a network of hollow veins, the number and distribution pattern of which remain absolutely constant for separate species, an arrangement suited to the insects and most helpful to the entomologist in classification and identification.

Although blood circulates through these veins during the insect's development, their main function is to provide strength, in the same way as the frame of an aircraft wing. In flight the maximum stress on the insect's wing is on the leading edge, so the veins on that edge are thicker and closer together. The longitudinal veins of the earliest insects were further strengthened by a network of crisscrossing veins between them, but the evolutionary tendency is to reduce the lateral veining and strengthen the longitudinal, at the same time improving the geometric pattern. The wing venation of the dragonfly still follows the old style, while that of the bee is thoroughly updated.

Unlike the wings of birds, these wing surfaces, whether of the bee, the dragonfly, or any other flying insect, have no muscular structure. They are merely superefficient aerodynamic surfaces controlled from the thorax of the insect. The thorax is where the power comes from, the only power. The wings of birds are modified forelimbs, and so have a muscular structure of their own; insect wings have none whatever. This is one of the reasons that the flight characteristics of insects are different from those of birds.

The thorax of the insect, to which the wings are attached, is a complex of flight muscles and mechanisms so utterly sophisticated as to boggle an aircraft designer's imagination. The thorax enables an insect in flight to carry out just about any maneuver, to loop, swoop, climb vertically, fly upside down, sideways, backwards, to hover, and to vary between all these in a fraction of a second. Watch a swarm of midges hanging in the air on a summer evening, hundreds and hundreds of them in a perpetual pattern of evasive action, never colliding, never falling, demonstrating flying techniques of breathless complexity.

That is how far evolution has taken the fragile-winged pests you recognize only for their nuisance value; the development, from amoebae, of probably a thousand million years, that makes our own ability to get up off all fours and stand and walk erect seem only a ludicrous gesture.

The way in which the power is applied from the thorax of the insect to the wings varies considerably from one order of insects to another. Basically the wings are hinged to the thorax via couplings which act like a series of ball-and-socket joints, so that they are free to move in any direction.

In the more advanced insects the thorax behaves like a box with walls stiff in some places and flexible in others. Deformation of its shape is effected by about ten pairs of muscles in each of the two thoracic segments that bear the wings. These power and control the flight.

For the majority of insects, the driving power for the wings comes from two pairs of muscles, the vertical and longitudinal indirect wing muscles. (Figure 1.) These are the largest in the insect's body and are not connected to the

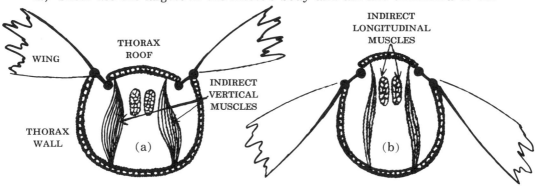

Figure 1: Diagrammatic transverse section of thorax, showing in (a) vertical muscles contracted, forcing wings up, and in (b) longitudinal muscles contracted, pulling wings down

wings but to the interior walls of the thorax. The vertical muscles run from the roof to the floor of the thorax, while the horizontal pair connects the front of the thorax to the rear. When the vertical muscles contract, the roof is pulled down and flattened. Due to the system of leverage, the wings move up. When the longitudinal muscles contract, the roof of the thorax arches upward, causing the wings to make a downward stroke. There are also some smaller indirect flight muscles which control the elasticity of the thorax by bracing its walls against the strain imposed on them by the main flight muscles.

As well as flapping up and down, the inclination of the wing has also to be varied during flight, so that it can be twisted along its axis. This is accomplished by the direct-flight muscles which are attached to the roots of the wing. Other direct-flight muscles effect the forward and backward motion.

The wings of dragonflies and some other more primitive insects are operated in a different way. In these, the two pairs are equally developed and are powered by large direct-flight muscles; one set pulls the wings up, and another set pulls the wings down. The muscles are directly attached to the wing bases, one each side of the pivot. (Figure 2.)

Such a primitive arrangement means that the four wings have separate power supplies, so that the two pairs need not beat in unison, but more or less independently. In the damselflies the front and rear pairs of wings can be out of phase by as much as one half cycle.

Many of the more advanced insects, such as flies, have what is known as a click mechanism, which increases the speed and efficiency of the wing strokes.

This involves modifications to a number of structures. In simple terms, the wing couplings are double-jointed, so that when the wings pass through their mid-position, the thorax walls are pushed outward before the wings can move fully up or down. These lateral movements of the thorax are resisted by accessory indirect muscles, until a degree of tension is attained. When this tension is released, the wings, instead of moving up or down smoothly, accelerate with a snapping action.

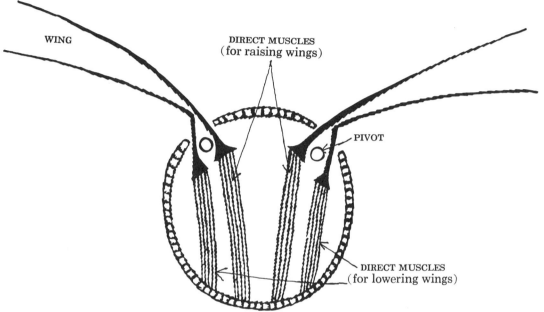

WING

DIRECT MUSCLES
(for raising wings)

PIVOT

DIRECT MUSCLES
(for lowering wings)

Figure 2: Flight mechanism of dragonfly

According to the need of the insect for its particular life pattern, Nature has devised an efficient flying mechanism best suited to that purpose. And in those cases where the wings of insects became a hindrance rather than an aid, Nature modified them in reverse, even to the point of total relegation.

Species from most orders are included in this grounding, including some moths, beetles, and even flies. Worker ants, for example, are born wingless, better for their life of underground toil, while the wings of the males and queens are used merely for the mating flight and are shed when this has been accomplished. Certain parasitic, bloodsucking flies have also lost their wings, which they hardly need when burrowing through the fur and feathers of mammals and birds.

In contrast, other insects have become so dependent upon their wings that they are virtually immobile without them. A fly crawls up a windowpane, but not so a dragonfly. It cannot walk at all and uses its wings as its only means of locomotion. But the exceptions and contradictions to the rules that govern the insect and its flight are few, and infinitely less interesting than the factors that make them fly.

In order to fly, an insect often moves its wings in a more complex way than a bird, and an awareness of simple aerodynamic principles is necessary to appreciate just how an insect flies and maneuvers through the air.

When an object is placed in a stream of air, certain forces act upon it. There is nothing surprising about the pushing effect of an airstream striking a surface, whatever its shape. Such a force can even produce a lifting effect on a flat plate if the plate meets the airstream at a slight angle. (Figure 3.)

AIRFLOW

Figure 3: Simple airlift on a flat surface

In this way a flat plate can be made to work as a simple wing and support itself against the force of gravity. However, the lifting and general aerodynamic characteristics of a wing are dramatically improved if it is shaped so as to produce maximum lift with minimum air resistance. Such a wing is called an airfoil.

The shape of an airfoil causes air to travel farther and faster as it flows over the top surface than the air flowing past the under surface. And when air moves faster its pressure drops, so that as a wing moves through the air there is less pressure on the upper surface than on the lower surface, and the wing is sucked upward. (Figure 4.)

Figure 4: Airflow around airfoil section

A number of other factors affect the lifting ability of a wing, such as shape, airspeed, and area of the wing. Lift obviously increases with airspeed and wing area. Air, however, resists the forward motion of a wing, exerting a force called drag which has to be overcome by increased thrust to move the wing through the air, or by gravity if the flight is descending. Flight depends on a balance of all these factors, and in level flight at uniform speed these factors will be in equilibrium. (Figure 5.)

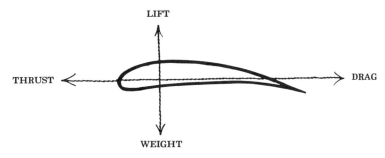

Figure 5: Forces acting on wing

Lift is further increased if the leading edge of the wing is raised at an angle to the airflow. This angle is called the "angle of attack." The larger the angle of attack, the greater the lifting properties of the wing, though this gain cannot be exploited indefinitely. Eventually a stage is reached when the smooth airflow on the upper surface breaks away and becomes turbulent. At this point, the "critical" angle, the wing, loses its lifting properties and stalls. (Figure 6.)

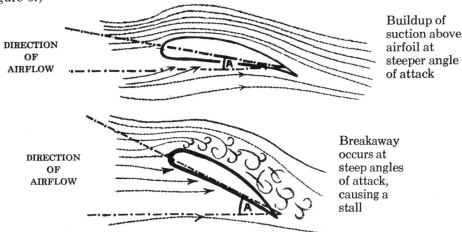

Figure 6: "Angle of attack," and "critical" angle in airfoil lift

The stalling angle for aircraft is about 15 degrees. If this is exceeded, lift diminishes dramatically and the plane falls like a dead weight. This is what makes landing a plane a precise maneuver, since in order to fly slowly the pilot requires a high angle of attack to compensate for the low airspeed.

The wings of insects, however, do not stall until they reach much higher angles of attack—up to 60 degrees in the fruit fly. And even when these extreme angles are reached, lift falls off gradually rather than suddenly. In this way insects receive ample warning to avert a stall. In fact, so advanced and automatic is the flight adjustment mechanism of most insects that they are incapable of falling from the air, enjoying a perfection of flying ability to make most pilots loop with envy.

Just how insect wings are capable of providing so much lift at high angles of attack is not altogether clear. Aircraft designers know that the aerodynamic properties of a wing depend on the behavior of the thin layer of air which is in contact with the surface of the wing. The way in which this boundary layer moves is critically important both in aircraft and insects. With insects, every scale, hair, bristle, corrugation, and vein plays a part in the wing's flying efficiency, although we know virtually nothing of the details of this efficiency.

Clearly, the supreme maneuverability of many insects cannot be explained in simple aerodynamic terms. Some scientists have even gone so far as to suggest that insects such as dragonflies and bumblebees should not be able to fly at all, but their calculations are based on the well-understood "steady-state" aerodynamics of textbook aeronautics. However, very recently Professor Torkel Weis-Fogh at Cambridge University has been pioneering fascinating research on insect flight, and has come up with surprising results which indicate that some of the established principles of aerodynamics are being apparently violated by the insect world.

He has discovered that to achieve lift, insects do not rely solely on ordinary airfoil action, but that the effects of "nonsteady" aerodynamics play a major role in their flight. He points out that as insects move their wings in an extremely complicated way, they produce fluctuating and unsteady airflow by means of a variety of novel aerodynamic mechanisms. The ingenious and complex explanations he puts forward make use of new terminology, such as the "fling and flip" mechanism, and the "clap."

A particularly significant aspect about nonsteady aerodynamics is that to achieve lift, the actual shape of the insect's wing is far less important than the creation of air circulation around the wing. This may provide a clue to the abnormally high angles of attack of which some insects are capable, and may even help to explain some of the bizarre wing shapes found within many orders of insects.

The wings of insects, because they are so thin and flat, do not resemble ordinary airfoils, although in fact they act as such. In addition, as soon as flapping starts and a flow of air passes around them, they change shape and become cambered into more efficient airfoils. (Figure 7.) This can be seen in many photographs throughout the book.

Figure 7: Wing of insect when flying

-24-

An airplane flies when the forward thrust provided by jet engines or propellers provides air flowing over its wing surface to produce lift. A helicopter is pulled upward by its whirling blades, but in an insect both propulsion and lift are provided by actual movement of the wings. In some respects insect flight is more like that of a helicopter, but insects' wings oscillate instead of rotate. The angle of attack alters at various phases of each wingbeat, and as the wings flap up and down they also twist on their axis, first one way, then the other.

Analysis of the movement of a fly's wing shows that it follows an ellipse or a figure eight, depending on the nature of its flight. In this way, the angle at which the front edge of the wing strikes the air is varied. During the down stroke the leading edge is lower than the hind edge, so that the relatively small angle of attack produces reasonable lift and little drag, but as the wing moves up again its attitude is practically vertical, with the leading edge above the hind. The combined effect of this action is to fan a current of air backward and downward, providing both lift and thrust.

The net result is that the insect moves forward and is simultaneously supported against the force of gravity. Unlike a powered aircraft, which uses its engines to provide thrust and its wings to produce lift, insects use their oscillating wings to perform both functions. This requires that the wing surfaces have to be twisted continuously throughout the wingbeat cycle, and such twisting can become so extensive that hovering, as well as backward flight, can be achieved. (Figure 8.)

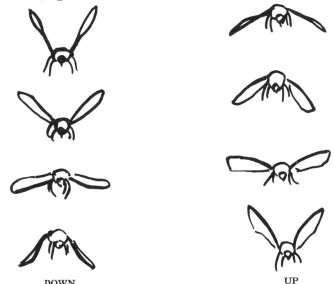

DOWN   UP

Figure 8: Wing twisting of fly at different instances of stroke

Though all are powered from the thorax, through a multiplicity of wing design of the same basic structured pattern, the performance of insects in the air varies enormously in terms of speed, endurance, and capability. It is, of course, keyed largely to the requirements of habit and habitat. Horseflies are very speedy, up to 30 mph, a trait no doubt required to catch a yearling on the home stretch. The fastest of all insects may be the large hawk moths, for they not only possess flight muscles more powerful than those of any other insect, but their superb streamlining adds to their speed potential. That they fly only at night compounds the difficulty of accurate measurement of their pace.

At the other end of the scale, midges, beating their wings up to twenty times faster than hawk moths, fly at one-tenth of the speed. The following table of wingbeats per second—one beat is the full up-and-down cycle—shows the surprising range of performance.

	Wingbeats per second	Meters per second
Medium butterflies	8—12	2—4
Damselflies	16	1—2
Scorpion flies	30	½
Large dragonflies	25—40	7—15
Cockchafer beetles	50	3
Hawk moths	50—90	5—?
Hoverflies	120	3—4
Bumblebees	130	3
Houseflies	200	2
Honeybees	225	2—3
Mosquitoes	600	under ½
Midges	1000	under ½

The amounts of energy needed by the flight muscles of the thorax to enable the wings to achieve these speeds is extremely high, so a directly convertible and easily replenished source of energy is required. Although fat serves as fuel for the majority of insects, butterflies and moths feed on nectar, a carbohydrate which they have to convert into fat for storage as their energy source. Bees and flies, on the other hand, use carbohydrates directly, in the form of sugars, without going through this need for storing energy. Both forms of fuel, fat and carbohydrates, have their merits and disadvantages.

The main advantage of carbohydrates is their solubility in water, allowing them to circulate in the body fluids which surround the muscles in the thorax, providing in this way instant sugar energy. In contrast, fat has to be broken down and transported from fat storage to muscles before it can be used.

However, the water solubility of sugar has certain disadvantages. Insects have only a small volume of body fluid, so they can dissolve only a small amount of sugar, and the potential supply of energy available in carbohydrates, weight for weight, is much less than fat. Blowflies, burning carbohydrates as fuel, may lose as much as 35 per cent of their body weight in an hour's flying, and bees run out of fuel after only fifteen minutes in flight. They must stop and fill up with nectar before continuing.

Fat, however, is so efficient as a fuel that in an hour's flight the migratory locust consumes only about one per cent of its body weight. Thus insects which have to sustain flight for long periods are the fat burners, while the busy, short-haul commuters rely on frequent sugar fill-ups.

Fueled by fat or carbohydrates, propelled by one pair of wings or two, flying fast or slow, across oceans or in elysian glades—only the slowest of the insects in flight, the large butterfly, provides a relatively clear image to the eye of man. The wingbeats of all others are a blur. We know what makes insects fly, we can see them moving through the air, but we have not been able to see precisely how they employ their wings for those daring and entrancing patterns of flight.

The common housefly, for example, can take off like a rocket, has directional control that can make it slip and slide through the air with derisive ease, and has the sensational ability to land upside down on the ceiling, from where it can contemplate its leaden-footed adversary.

Scientists have variously claimed that to achieve this landing a fly rolls before it settles, or that it loops just before touching down. The problem has baffled man through the centuries since he first watched a fly on the roof of his cave.

Now he need surmise no more, for the photographic appendix to this book shows quite clearly what happens, and the truth turns out to be spectacularly simple in concept, yet dependent on most sophisticated flight control for its completion. The fly flies straight up toward the ceiling, uses its two front feet as touchdown plates, then virtually cartwheels over until its four other feet complete the soft landing.

Though this determination was hardly the main purpose of several years of photographic research, I am nonetheless proud to have recorded the event in a photograph of an English kitchen ceiling. It was one of several fascinating findings on the flight of insects revealed by the camera and shown in the illustrations to this book.

ODONATA

DRAGONFLIES AND DAMSELFLIES

Tennyson's poetic imagery of the dragonfly, his wings "clear plates of sapphire mail," may be an entomological overstatement, but it does reflect accurately man's almost hypnotic enchantment with these elegant creatures of light and air.

Gently threading their bright colors through the sunlight, dragonflies are an adornment to our world and to the cultural lore of many lands. They have a quality of mystery that intrigues the senses.

In Japan dragonflies are associated historically with victory in battle, their likeness depicted in art, their importance recorded in literature. To Hopi Indians, living in desert lands of the Southwest United States, dragonflies became religious symbols of life, their design richly stylized on ceremonial pottery.

The largest dragonflies of all—a South American species with a seven-inch wingspan—enjoy an eerie distinction. At dusk, the pale wing tips of these twilight fliers, glinting in gloomy jungle glades, take on the form of ghostly, luminous windmills, regarded fearfully by superstitious tribes as the embodiment of departed human spirits.

Cackling witches of medieval England intoned that the long, needlelike bodies of dragonflies would sew up the lips of naughty, lying children. And many Asians welcome dragonflies as food, crisping them to a turn in the family wok. Yet, whether we treasure their nutritional value or imbue them with ancestral portent, one thing is certain. We are aware of them. They weave their spell.

Dragonflies are benign in their attitude toward humans. They neither sting nor bite us. They do not devour crops or spread disease. And they possess one unique fact beyond their scintillating brilliance: their vision of the ages. For they have watched the changing scenes of life on earth through well over 300 million years.

Before the era of the early dinosaurs, giant dragonflies with a wingspan of up to thirty-six inches—the largest flying insects of all times—hovered over tropical lagoons where eighty-ton, forty-feet-tall *Brachiosaurus*, the largest land animal ever, wallowed in primeval mud. But gargantuan scale was not an imperative of life in the early slime, for other fossils give evidence of smaller species, much like their shimmering descendants today.

Dragonflies survived the violent upheavals of the mountain ranges and witnessed the great mammal extinctions whose mystery continues to baffle detective scientists. Dragonflies were already incredibly ancient when, a mere two million years ago, grunting, stumbling man first watched their iridescent flight. Dragonflies have seen the newcomer clothe himself, arm himself, and begin the deliberate extermination of other animals around him. Aloof and shining, *Odonata* may well also witness the brief intruder exhaust his tenure of this earth.

Odonata is the oldest surviving order of flying insects, and, as an additional distinction, the aerial equipment of the dragonfly has remained essentially unchanged. Yet, antique though the design may be, *Odonata* share one thing with true flies, the most advanced of nature's fliers. Dragonflies and true flies both employ a built-in flight-stabilizing mechanism, though each by means of an entirely different system. As far as we know, none of the intermediate orders of flying insects has succeeded in this.

Whereas the fly has developed a sophisticated, gyroscope-like method of reassuming its flight path, the dragonfly actually uses its head for the same purpose. The head is connected to the body in a way that allows very free movement. Differences in position of the wings and body relative to the head, which maintains level attitude, are detected by four beds of hairs, between the head and body. Through this flight-information system, the insect is continually made aware of its "attitude" and compensates as necessary. (Figure 9.)

Despite their aerial prowess, dragonflies enjoy only a brief life of beauty, usually just a few weeks on the wing. The remainder of their total lifespan of two years or more is spent as unlovely underwater larvae—fat, squat, spiny-legged—scouring riverbeds and ponds. These larvae are aggressive, formidable aquatic predators, their protruding eyes searching for water insects, which they can snatch at great speed, with jetlike propulsion, in their hungry, thrusting mouths. The larvae of larger dragonflies even add small fish to their menu.

Like other more primitive insects, dragonflies have an incomplete metamorphosis. They undergo no dormant, chrysalis stage, as do more advanced insects such as moths and butterflies. Instead, they lead an active existence

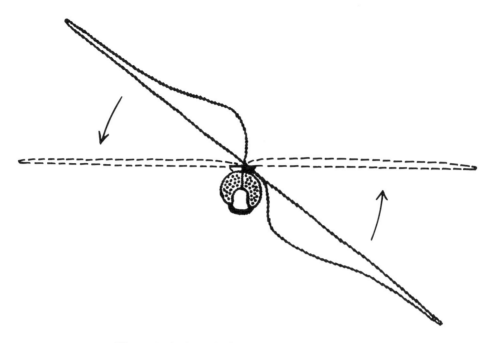

Figure 9: Action of wings when body is displaced

throughout, molting two or three times while in larval stage, each time shedding a complete skin to accommodate their lengthening bodies and embryonic wings.

When the dragonfly finally takes to the air—a complete contradiction of its former existence—it is as well able to fend for itself as it was underwater. The extraordinary eyes can see as far as forty yards, almost in a complete circle, enabling it both to detect enemies and scan for food. Particularly partial to mosquitoes, the dragonfly grabs them in a cage formed by its four legs bunched in front of the wings. Once snagged, the morsels are deftly transferred from cage to mouth, and the aerial gourmet speeds on, often flying as fast as thirty miles per hour.

As though living nine-tenths of its life in another element were not traumatic enough, the dragonfly further astounds us with a mating procedure that would make a less determined insect consciously avoid regeneration of its species. The male has two hooks on his tail with which he grabs the female around her neck. She responds by swinging her body up under his, forming a wheel. With this spectacular and acrobatic embrace, they mate in flight. The act is completed when the female swoops across the surface of some tranquil pond, allowing the water to wash off her fertilized eggs, where they sink to the bottom and begin their slow life cycle.

The order *Odonata* also includes damselflies, which are generally smaller than dragonflies and have patterns of behavior that set them slightly apart from their larger cousins. They are nonetheless true dragonflies entomologically, in all but the name given them for convenient distinction.

Damselflies employ a slower, more flickering flight style. They do not eat in the air, preferring to search for insects on foliage, and they are able to fold their wings over their bodies, unlike the dragonflies, whose wings always remain spread out in a flat, horizontal position. Damselflies mate in the same gymnastic fashion as dragonflies, though not in flight, and the female often deposits her eggs in the folds of some aquatic plant, where her larvae lead a less active underwater existence than do the free-roaming offspring of the dragonfly.

As the softness of their name suggests, damselflies are more gentle in manner than dragonflies. They are less aggressive in their territorial defenses, populating an area more densely, living together more amicably. They are more hesitant and less showy, but nonetheless attractive in their shy way.

Dragonflies and damselflies comprise a most beautiful order of insects, their several thousand variably colored species flying wherever inland waters catch the glint of the sun. More than three hundred years ago, Thomas Mouffet wrote that the dragonfly "doth set forth Nature's elegancy beyond the expression of Art."

If he were alive today, Mouffet might well add that the camera has now contributed to the fragile allure of earth's most ethereal fliers.

Where glowing embers through the room
Teach light to counterfeit a gloom,
Far from all resort of mirth,
Save the cricket on the hearth.

—*John Milton: "Il Penseroso"*

ORTHOPTERA

GRASSHOPPERS AND CRICKETS

By and large, humans tend to feel affection for animals whose characteristics and behavior they can relate to their own. We like animals we can stroke, those whose expressions seem human, those which reciprocate the affection we lavish on them, and most of all those which become dependent upon us.

Insects, however harmless, are much too remote. A gnat cannot easily be stroked, except presumably by another gnat, and it certainly won't come to be fed when it's called, retrieve a stick, bring your slippers, or covet the warmth of the fire—unless it's a cricket on the hearth.

Which makes grasshoppers and crickets—sometimes referred to as merry —an exception to the rule. They do come closer to a degree of affinity with us, cheerful, chirpy little jumpers that they are.

Grasshoppers, crickets, and their near relatives, mantids and cockroaches, are members of the order *Orthoptera*, the "straight winged." Most orthopterous insects possess long, narrow, leathery forewings in camouflage hues of browns and greens. In contrast, their hindwings are broad, membranous, and netlike, and in many species often boldly colored red, orange, or yellow. When not in use, the hindwings are folded in pleats over the body, and the forewings are positioned over the hindwings, serving as protective covers.

Although some species of *Orthoptera*, notably locusts, are capable of undertaking long migratory flights, the majority are reluctant to use their wings, usually restricting themselves to very short distances, at best. In grasshoppers and crickets, take-off is aided by thrust from the large and powerful hind legs, which have been specially developed for jumping. Some species have even supplanted wing use altogether, relying only on super hind legs for propulsion.

Grasshoppers and locusts are big eaters, and most are strict vegetarians, unlike the voracious, predatory mantids which trap other insects, and even, in some large species, small frogs and lizards, with their highly specialized front legs. Yet it is the plant eaters, not the carnivores, which are the great threat to man.

The large, chomping jaws of the locust satisfy a hearty appetite that goes relatively unnoticed when he is leading his solitary, independent existence. But when food is scarce he tends to congregate, and in extreme conditions migrates in dense formation. Deprived of food, the locust actually grows longer wings and longer hind legs, equipping himself for his desperate search for food.

Millions of humans have died of famine induced by locusts, winging in dense black clouds across rainless skies, gobbling the crops that nurture man, stripping the sustenance of life before his unbelieving eyes, then vanishing to become a dread legend to be feared again. But scientists have now determined a pattern of behavior in these marauders that suggests that they are themselves victims of Nature's imbalance, not some mystic retaliation for man's transgressions.

Locusts are short-horned grasshoppers which become strongly gregarious when faced with starvation, migrating as a last resort for self-preservation. Provided they have sufficient food in their home territory, they lead contented lives that pose no threat.

Grasshoppers and crickets, together with cicadas (which belong to another order) are great musical instrumentalists, producing those lulling sounds of summer by rubbing projections on the under surfaces of their upper wings against a rough area on the top surfaces of their under wings. The vibrations from the wing surfaces are then amplified through the wing membranes. Only the males orchestrate, and the music is their love call. Species can even be identified by their distinctive songs.

The chirping rate of crickets is much affected by air temperature, accelerating as the temperature rises. In the snowy tree cricket, for example, it is possible to gauge the temperature quite accurately by applying a simple formula:

$$\text{Temperature } °F = 50 + \frac{N-40}{4}$$

where N is the number of chirps per minute. An engrossing way to spend an hour or two on a summer's day.

In Japan, Portugal, and Italy grasshoppers were kept in cages, though more for the value of their soothing song than as thermometers. In his *Natural History of Selbourne*, 1789, the English naturalist Gilbert White wrote that a field cricket "when confined in a paper cage and set in the sun, and supplied with plants moistened with water, will feed and thrive, and become so merry and loud as to be irksome in the same room where a person is sitting." And presumably vice versa.

. . . this bug with gilded wings,
This painted child of dirt, that stinks and stings.

—Alexander Pope

HEMIPTERA

BUGS

Though there are many delightful exceptions, it is difficult, as Alexander Pope found, to be enthusiastic over the majority of *Hemiptera*, the true entomological bugs. Living on the sap and fluids of plants and animals, many transmit disease and wipe out crops and herds. Among them, only the cicada atones for its gluttony, singing to us merrily as it cuts down life.

Bedbugs are probably the most notorious of the tribe, and though numbering no more than twenty species worldwide, they have an established reputation. Sadly, however, they do not possess wings, so it was not possible to photograph them in flight beneath the covers.

Unacceptable though their habits may be, bugs are, nevertheless, a fact of life, comprising a large collection of apparently dissimilar insects of about 50,000 species, divided into two groups. Members of the first suborder, the *Heteroptera*, have normal membranous hindwings, but unusual forewings in that the basal half is thick and leathery (as in grasshoppers), while the terminal portion is thin and transparent. *Heteroptera* include plant bugs, shield and stink bugs, assassin bugs, and water boatmen. As in most orders, not all the members have functional wings, but those that have, fold them flat over their bodies when not in use.

Bugs in which the structure of the forewings is usually uniform with the hindwings are placed in another suborder, the *Homoptera*. Included in this group are the cicadas, froghoppers, and aphids, distinguished by the manner in which they fold their wings roof-like over the sides of their bodies when at rest.

Bugs of both the subgroups share in common the long, beak-like mouthparts used for piercing plants or animals and sucking juices or blood. They are also capable of injecting salivary fluids into the tissues of their host, which has a digestive and liquefying action, aiding ingestion.

Though *Hemiptera* are mainly inimical toward man, in the sense that they eat his harvests, there are among them several unusual, even charming, species. The rhododendron leafhopper, common throughout North America and first introduced into England in 1935, launches itself into the air with a leap and then opens its wings to complete the flight. For rhododendron leafhoppers this dual flight style works perfectly as they skip from leaf to leaf, sucking the juice from the rhododendron plant.

Another bug with a dual style is the water strider, and still another the water boatman. The first rarely uses its wings, preferring instead to take part in those spirited regattas you can watch on the surface of almost any pond. Sculling away enthusiastically, it does not actually break the skin of the water, and seems to revel in its nonchalant aquatic skill.

The water boatman also has wings, but spends its time beneath the surface, usually finding it unnecessary to fly. These two bugs interest and amuse me, and I am tempted to wait interminably by my favorite pond on the off chance that a strider will take off like a seaplane, or a boatman emerge from the deep on its vertical flight path. Nature, in her own time, may well trim the wings from these reluctant fliers.

Certain bugs do provide material for our use, in the preparation of dyes from the mermis and cochineal bugs, and of wax from the Chinese wax insects. The resinous secretion of the lac insect is used diversely in shoe polishes, varnishes (lacquer), wood filler, lithographic ink, and even in the glazes on confections and coffee beans.

But this assortment of luxurious trivia hardly compensates for the mountainous harvests that *Hemiptera* take from man.

Now lead me hence,
And in a gilded bower
Let lacewings weave my dreams.

—English, Anonymous

NEUROPTERA AND MECOPTERA
LACEWINGS AND SCORPION FLIES

The transparent wings, iridescent bodies, and long, thread-like antennae of the lacewings make them the epitome of aerial grace. No flying insects look more fragile and delicate, yet the golden eyes and soft, pale wings of the slender green lacewing, for example, are only the external finery of a voracious, active insect well able to fend for itself.

The weak, fluttering flight of these bewitching creatures appears haphazard and uncontrolled, at the mercy of the merest puff of wind, yet again, like their fragile appearance, this first impression hides a more efficient reality. They are flying machines capable of sensational loops and vertical take-off, as I finally proved after experimenting with more than 900 exposures.

Green lacewings are found in most parts of the world, some species in England being virtually identical with their North American cousins, only their stronger odor setting them somewhat apart. Unlike many insects, adult lacewings survive the winter in hibernation. In the autumn they often seek shelter indoors, their bodies slowly changing to an orange hue, then turning green once more to greet the spring.

Lacewings are part of an order, *Neuroptera* (nerve wings), comprising more than 4000 species, and including alderflies, dobsonflies, antlions, and snake flies. All *Neuroptera* have four large, membranous wings of equal size which close over their bodies when at rest, like tents, but the different families vary their life-style and appearance quite distinctively.

Alderflies have smoky-colored wings and frequent the vegetation bordering ponds and streams. Their larvae are aquatic and possess several pairs of long segmented gills on each side of their abdomens. Dobsonflies, though similar in habits, are generally much larger, some species in North America and New Zealand with a wingspan of more than four inches. Male dobsonflies have great pincer-like jaws out of all proportion to the rest of their bodies.

Antlions bear some general similarity to dragonflies, but have knobbed antennae. Snake flies are similar to lacewings, but can be distinguished by the long neck, which is really an extension of the thorax. All four have in common those double pairs of large, flimsy wings.

Scorpion flies are members of a much smaller order, *Mecoptera*, and can be recognized by their prominent downward-pointing beaks. Their four wings are of approximately the same size as those of *Neuroptera*, but although transparent, they are often marked with brown or yellow patches. As in the lacewing, the antennae of the scorpion fly are long and thread-like.

Scorpion flies frequent damp, shady woodlands over most of Britain, Europe, and North America, and can usually be seen flying around low vegetation. They have an almost indiscriminate appetite, moving from plant to plant in search of dead or wounded insects and feeding on nectar and oozing sap. Occasionally they will attack perfectly healthy, smaller insects.

On the whole, they are not aggressive. However, members of one family of scorpion flies look like large crane flies and catch their prey by hanging onto a leaf or twig with their front feet, dangling their long hind legs in space, ready to seize any small insect unfortunate enough to fly within reach.

The scorpion fly earned its name from the upturned tip of the male's abdomen, which looks very much like the sting of the scorpion, though it carries no sting and is used only for mating.

The flight of lacewings and scorpion flies is particularly fascinating. Their waywardness in front of the camera drives me nearly to distraction, and capturing their elusive beauty is always a challenge.

*I do not know whether I was then a man
dreaming I was a butterfly, or whether
I am now a butterfly dreaming I am a man.*

*—Chuang Tzu,
Chinese philosopher, 300* B.C.

LEPIDOPTERA

BUTTERFLIES AND MOTHS

In American Indian legend the Great Spirit breathed life into the myriad colored pebbles of the shimmering streams and gave them wings so that they could display the gentle beauties of nature to his people. In this way butterflies were created for the delight of mankind, bright symbols of velvet delicacy, hovering wherever flowers bloom.

Entomologically, butterflies and moths comprise the second largest order of insects, *Lepidoptera*. More than 100,000 species have so far been identified, inhabiting every land mass except Antarctica. They range from the smallest micro-moth, with a wingspan of less than one-eighth of an inch, to the giant owlet moth of South American jungles with a wingspan of up to twelve inches.

Butterflies and moths have so many characteristics in common that there are no foolproof ways to separate them within their order, although in most cases there are obvious differences. Butterflies close their wings to a vertical position, while the majority of moths fold them flat across their backs when at rest. The antennae of most butterflies terminate in small knobs, while those of moths are usually feathered or serrated. And when it comes to flying, most butterflies employ a slower, more flickering style, traveling during daylight, while moths are generally much faster fliers, hurrying through the night in search of food and mates.

In common, they all have two pairs of large and often colorful wings, the surfaces covered with thousands of minute scales overlapping like tiles on a roof. The bright, distinctive, and often intricate patterns on the wings of *Lepidoptera* are made up of a mosaic of these tiny units. The wing membrane itself, beneath the scales, is transparent. The bodies, too, are similarly clothed, but with longer and more hair-like scales.

In butterflies and moths each pair of wings is coupled together during flight, making two large and more efficient airfoils. Most moths have one or more bristles on the base of their hindwings (frenulum) which hook onto a special catch on the underside of their forewings (retinaculum). Butterflies have a lobe on the leading edge of the hindwing which locks under the overlapping forewing. (Figure 10.)

If in many ways they are similar entomologically, butterflies and moths are regarded most disparately by the casual observer. How can the shy charmer that dances so daintily for us in sunlight be cousin to the persistent flutterer

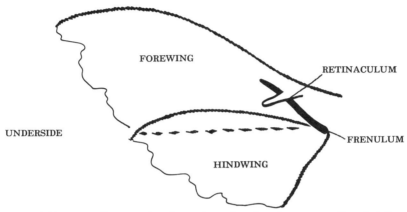

Figure 10: Wing-coupling mechanism of moths showing connecting bristle (frenulum) and catch (retinaculum)

that plagues our summer nights? Little wonder that moths were held to be the souls of the dead come back to haunt us, while butterflies transport our spirits heavenward. The flying pebbles of the Great Spirit seem infinitely preferable.

The mouthparts of both the butterfly and the moth consist of a highly developed long proboscis, used for sucking nectar and other liquids. When not in use it is coiled up like a watch spring beneath the head. A few moths, however, such as silk moths, have only vestigial mouthparts and are unable to feed at all. Instead they exist solely on the food stored in the fatty tissues from their caterpillar days.

The butterfly has a sensational taste mechanism, located on the soles of its feet. As soon as the insect's feet detect the sweetness on the petal of a flower, an automatic reflex uncoils the tongue and makes it probe into the nectary. Monarch butterflies are said to react in this way to a solution containing as little as one part of sugar to 120 million parts of water.

The sense of smell in butterflies and moths is transmitted through their antennae, which in the female are also used for identifying suitable kinds of plants on which to lay her eggs. In male moths the main function of the antennae is to track down the females, usually in total darkness. The antennae of some male moths are incredibly intricate, divided and subdivided into numerous branches forming large plumy structures. In this way a large surface is in contact with the air. In species such as the Emperor moth, the antennae are so sensitive that they can detect and home in on the female moth scent several miles away.

There is often an astonishing difference in the appearance of male and female butterflies and moths of the same species, a variation sometimes so great that the male and female look like separate species altogether. In butterflies the most spectacular difference occurs in the birdwings from the islands of Indonesia. The males of these beautiful insects are not only very much smaller, but are utterly different in color, pattern, and even shape.

In moths the most striking contrast of all occurs in species such as the European winter moth, *Operophtera brumata*, and the American fall cankerworm, *Alsophila pometaria*, in which the female is flightless, her wings either totally absent or reduced to mere vestiges.

Of all orders of insects, butterflies and moths exhibit the widest range of flying characteristics and abilities. Many of the butterflies, such as Monarchs, Red Admirals, and Painted Ladies, are strong, fast fliers, able to migrate thousands of miles, even across the Atlantic; whereas some of the white butterflies of the Pieridae family, by comparison, seem hard pressed to reach the next flower. The differences are equally marked with moths, from the powerful Hawk to the weakly fluttering Plume and Footman moths.

Of all insects, butterflies appeal most to our poetic senses. They have romantic connotations of Victorian girls in white muslin dresses running decorously through golden meadows, their nets raised to scoop a tiny portion of heaven on the wing. Blackfoot Indians even believed that their dreams were brought to them by butterflies. With evidence like this, can there be doubt that butterflies are messengers from paradise?

A bell tower to a beetle
Is a challenge of degree
The banquet uncompleted
'Til the timbers are depleted.

—English Ode

COLEOPTERA

BEETLES

As Christopher Robin discovered to his dismay, the trouble with beetles is that they tend to escape from matchboxes.

> We went to all the places which a beetle might be near,
> And we made the sort of noises which a beetle likes to hear,
> And I saw a kind of something, and I gave a sort of shout:
> "A beetle-house and Alexander Beetle coming out!"

Christopher Robin was a lucky boy, for the incidence of lost beetles recovered is not high. Their ability to disappear into the merest fissure, at once beyond recall, accounts for many a loss, though the relative ease of substitution usually makes search unnecessary. There are so many more.

Entomologists, in fact, will readily quote that *Coleoptera* outnumber in species all other orders of living creatures, that there are at least five times as many beetles as there are all vertebrate animals combined, and that estimates of beetle species so far classified exceed 210,000, with possibly as many more still to be registered.

There are so many beetles in so many places that a beetle phobiac can escape them only in seawater. For, excepting the oceans, beetles have invaded practically every environment of our planet, from the bleakest to the hottest, from the most exotic to the coldest. Their diverse habitats include wood, carrion, dung, vegetable matter, fungi, freshwater. And their profusive ways make them, by any standards, a most successful order.

The general public does not warm to them as an order, however, though their number-one charmer, the ladybug, is beloved by all, a harbinger of luck when she alights upon us, then quickly departs to resume the search for her errant children. Fireflies, too, dancing luminously in the dark of summer nights, never fail to delight us. Both kinds may seem unlike their black-backed cousins, yet all are true beetles.

Coleoptera do little toward humans in direct affront, but behind the scenes they favor a diet of cathedral roofs, ships' timbers, and wooden forts, with crops and clothing for supplementary diversion.

To a great extent the success of beetles is attributable to their defensive armor. They are protected by an exceptionally hard and thick exoskeleton (insects do not have skeletons like vertebrates; instead their bodies are supported and protected by a shell-like external coat of chitin known as an exoskeleton). This coat of armor is further strengthened by the modified forewings which have been transformed into tough and horny impermeable covers called elytra. These protect not only the body of the beetle but also the delicate membranous hindwings intricately folded beneath.

Most species of beetles take to the air only as a last resort. They have largely become earthbound, being slow and clumsy on the wing, unable to make immediate or emergency take-offs, or take swift, evasive action when airborne. When they do decide to fly, there is a long delay, sometimes of several seconds, before they launch themselves. During this time the elytra have to be raised, and the wings, which are folded crosswise and lengthwise in an extremely complicated way, have to be unraveled and spread out.

Finally airborne, they fly like ponderous freighters, lacking directional stability, veering to one side or the other uncontrollably, often hitting objects and stalling. In flight, the elytra are held at about 45 degrees above the body, functioning in a limited capacity as fixed airfoils, and increasing lift. The stroke angle of beetles is exceptionally large, in many species extending to about 180 degrees—sometimes making the wings even touch below the body.

Beetles have been regarded through the ages as beneficial medicinally, particularly when powdered. And to ancient Egyptians, in the form of the scarab, they were regarded as symbols of eternal life.

To the entomologist, beetles comprise a most engrossing order of insects, their fascination shared by few others except poets. William Shakespeare had a certain regard for them, reminding us in *Measure for Measure:*

> And the poor beetle, that we tread upon,
> In corporal sufferance finds a pang as great
> As when a giant dies.

Which, for a cold-blooded insect, is probably not strictly true, though the thought remains valid as a standpoint from which to regard our fellow creatures.

There's a whisper down the field where the year has shot her yield,
And the ricks stand grey to the sun,
Singing:—"Over then, come over, for the bee has quit the clover,
And your English summer's done."

—Kipling: "The Long Trail"

HYMENOPTERA

BEES, WASPS, AND ANTS

In my tender years I was led to understand that bees made honey and wasps made marmalade. Erroneous the latter, as I soon discovered, but not so far-fetched to a young believer when you consider that both appear in glass jars on the breakfast table.

Apart from that intriguing possibility, wasps do not normally excite us, except in our heed to escape them, but we do like to be associated with the more commendable attributes of bees and ants. They are the epithetic insects, "busy," "toiling," and lauded by poets and philosophers through the ages. They hold a particular fascination, these workers of the insect world, dazzling us with their industry and their social orderliness.

According to the ancients, honeybees were sent from Paradise as an example of desirable behavior to frail and fractious mortals. They continue to demonstrate their impeccable social harmony, largely to an unheeding world, at the same time converting the flowers of summer into a food favored by heaven and earth alike. Jupiter, lord of the heavens, Roman god of victory and light, was nurtured by honeybees, and earthlings have since attributed many supernatural properties to nature's golden elixir.

Yet, though we extol the virtuous bee, fortunately we cannot emulate this winged paragon, for he is merely a computerized automaton, incapable of the behavioral errors which make fallible man so fascinating.

Bees and ants are members of the order *Hymenoptera*, in which they are joined by wasps and a myriad array of sawflies, parasites, and ichneumon flies. They total more than 100,000 species so far identified, in every location from arctic tundra to tropic jungle. And they show a staggering diversity and specialization of habit, ranging from the highly developed caste societies of honeybees to the unenviable life-style of the larvae of certain parasitic

– 43 –

wasps, able to survive only within the parasites of other parasites. Complex, entomologically fascinating, and rarely claimed as candidates for human comparison.

In yet another species, *Hydrophylax aquivolans*, the female dives underwater and uses her wings to swim, like a penguin, in search of damselfly eggs, in which to lay her own. Despite this appealing aquatic performance, honeybees, of course, remain man's favorite among *Hymenoptera*, their merits firmly established in the legends and folklore of the world.

Nero's wife, Poppea, ascribed the sheen of her flawless body to a balm of honey and wild asses' milk. Napoleon ordered an embroidery of golden bees upon his imperial robes, as yet another reminder of his God-given power. Medieval old-wives' tales of the efficacy of bee stings as an antidote to rheumatism still count many firm believers, and if the remedy could be rendered painless, no doubt its followers would be as numerous as today's wearers of anti-arthritic copper bracelets.

Though generally unappreciated in Anglo-Saxon lore, in many parts of the world ants, too, are highly prized both for their nutritional and medicinal value. The supposed wisdom and skill of ants caused Arab parents in times past to place one in the hand of a newborn infant, with a prayer that the baby would develop similar attributes, and in Guiana a particularly vicious breed of ants was encouraged to bite the feet of crawling children, as a practical way to make them walk sooner.

Medically, the soldier ant is used by Amazonian jungle tribes in the closing of deep flesh wounds. Placed on an incision, the sharp, saw-toothed jaws of these ants, held at intervals along the wound, effect a very satisfactory suture, the firm grip remaining even when the heads of the ants are pinched off. This ingenious method of stitching was also much used in countries bordering the Mediterranean, where carpenter ants were employed for suturing before the spread of more modern surgical methods.

Ants even have value as a dietary supplement. Honey ants store honey in their abdomens, which swell into huge, bulbous, immobile food banks. Confined to the nest, the only function of these storage ants is to serve as food reservoirs, to be tapped by the workers between forays on their endless tasks. Tribes in Central America and aboriginals in Australia prize these honey ants as a delicacy, cheerfully raiding their nests and squeezing the honey sacs directly into their mouths. And in some Asian countries ants have long been regarded as a welcome treat—fried, dried, and even "on the hoof."

In typical *Hymenoptera*, such as wasps and honeybees, each insect has two pairs of transparent wings, which in flight are coupled together in an ingenious zip-fastener arrangement. A row of tiny hooks on the leading edge of the hindwing engages with a fold on the trailing edge of the forewings, creating a single, large, more efficient wing area. (Figure 11.)

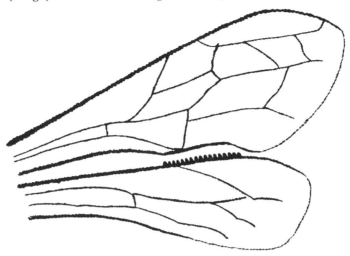

Figure 11: Wing of honeybee showing row of hooks on leading edge of hindwing for coupling wings in flight

The obvious characteristic of most members of the order is the "wasp waist," though sawflies and wood wasps can be distinguished from bees, wasps, and ants by having no narrow waist between the thorax and the abdomen. These unwaisted species, incidentally, do not sting, a characteristic which makes them rank lower in the order in an entomological sense, though rather higher in popular esteem.

Hymenoptera are essential to the success of man's environment, perhaps more so than any other order of insects. In their ceaseless underground foragings, for instance, ants aerate the soil thoroughly and efficiently, while bees play a crucial role in the pollination of fruit crops and plants, and provide a rich, natural food source. Some biologists even go so far as to claim that without the role of parasitic wasps in regulating the population of plant-feeding insects, the higher mammals would never have evolved.

We may tend to overstate their importance, but certainly it is true that without this busy order of insects our world would be a very different place. As it is, we applaud the purposeful bee, speeding toward its hive with another load of sweet nectar. He is doing good work.

We appreciate bees. They are linked inescapably in our minds with the drowsy peace of a summer garden. Their humming as they go their busy way suggests a contentment, a reassurance that nature is on course. Perhaps they are indeed doing heaven's work here on earth.

Go, poor devil, get thee gone!
Why should I hurt thee? This world is
surely wide enough to hold both thee and me.

—*Laurence Sterne:* Tristram Shandy

DIPTERA

TRUE FLIES

Laurence Sterne was unusually sympathetic in his attitude toward flies, for most of humankind quite actively dislikes them, brandishing aerosol cans across the world with furious abandon. Harried unmercifully, flies are blamed for the spread of disease, are associated with filth generally, and boast few champions.

Not surprisingly, the rate of attrition is high, but the regenerative capacity of flies is also high, so they persist. Which is just as well, for they play an absolutely essential role in the balance of nature, and without them we would be hard put to survive.

Flies are insect scavengers, the undertakers of the world. The dead, whether animal or vegetable, are consumed by their larvae. Denied their industry, the surface of this earth would be more untidy than even man has managed to make it. How helpful if they could also eat junked cars.

The removal of debris is only one of the tasks they perform. Predatory and parasitic flies feed on many of the insects that are harmful to the crops of man, and flies play a large part in pollination. Look at any patch of wildflowers on a summer's day, and you will see more flies than bees feeding on them and pollinating them.

True flies, or *Diptera* (two-winged), comprise one of the four large orders of insects, with more than 60,000 known species. Although the typical fly can be easily identified as such, some, owing to their shape and coloration, can readily be mistaken for bees or wasps, which they can at times mimic so well that even an entomologist is confused at first glance.

The main distinguishing feature of flies is their single pair of wings. The more advanced groups of insects have shown an evolutionary trend to reduce their two pairs of wings to a single pair. Some have developed mechanisms for coupling their forewings and hindwings together in flight, but the true flies have discarded their hindwings, and all that remains of them is a pair of knobbed stalks known as halteres.

At one time it was thought that these halteres acted as balancing organs, for flies deprived of them go completely out of control, but recent research shows that halteres have a gyroscope-like action more akin to the "turn and slip" indicator of an aircraft.

When the wings are in motion, the halteres vibrate through their horizontal and vertical axes, so that during normal flight symmetrical impulses are sent to complex sense organs at their bases. Any change in the flight attitude affects the regular oscillations of the halteres, and the fly is subsequently "informed" of its new condition. In this way they perform a function similar to that of the head of the dragonfly.

How do flies achieve such brilliant performance in the air, accelerating, looping, rolling, able in an instant to change their flight paths? In larger insects with a relatively slow wingbeat, the muscles powering the wings are controlled by separate nervous impulses for each wingbeat. But in the case of smaller insects, such as flies and bees, requiring much faster wingbeats to keep them airborne, it is impossible to control the wings in this way, for conventional muscles are quite unable to contract and recover in time for each new phase of movement.

The fast fliers, therefore, evolved a revolutionary type of muscle tissue known as fibrillar muscle. It is the most active tissue ever devised by a living organism, with the peculiar ability to contract automatically and rapidly when stretched, then immediately to relax after each contraction, enabling the wings to be operated at quite incredible speeds.

Development of this fibrillar muscle is related to the "click mechanism." When the longitudinal muscles of the thorax contract to move the wings downward, the wings click over the midpoint, suddenly releasing the longitudinal muscles from tension and simultaneously giving the vertical muscles a quick tug. Automatically the longitudinal relax and the vertical contract, pulling the wings up again. As soon as the wings click over midposition once more, the opposite action takes place. Once an impulse from the central nervous system starts this mechanism, the main power supply for the wings runs automatically until another nervous impulse stops it. In this way the frequency of the wingbeat is not directly under the control of the insect but depends on various physical factors, such as the size of the insect's wings and the surrounding air temperature.

The development of the click mechanism reaches its peak in true flies and in *Hymenoptera*, the two orders with higher wingbeat frequencies than those of all other insects. The mechanism is a highly evolved one requiring special muscle tissue combined with an elaborate articulation of the wings. It is this combination that enables the smaller and more advanced insects to beat their wings at such very high speeds.

Houseflies have a wingbeat frequency of about 200 per second. Some midges even go as high as 1000, though there are peculiar aerodynamic problems at these speeds that make the normal properties of airfoils change. In these conditions the insect is not flying in an aerodynamic sense at all, but rowing its way through the air—at 1000 strokes a second. A very sophisticated method of propulsion indeed.

Next time you try to swat a fly and miss it, console yourself that your adversary is employing Nature's most advanced flying techniques to avoid you. Hundreds of millions of years of evolution have contributed to your momentary frustration.

Plate 1

Plate 2

Plate 3

Plate 4

Plate 5

Plate 6

Plate 7

Plate 8

Plate 9

Plate 10

Plate 11

Plate 12

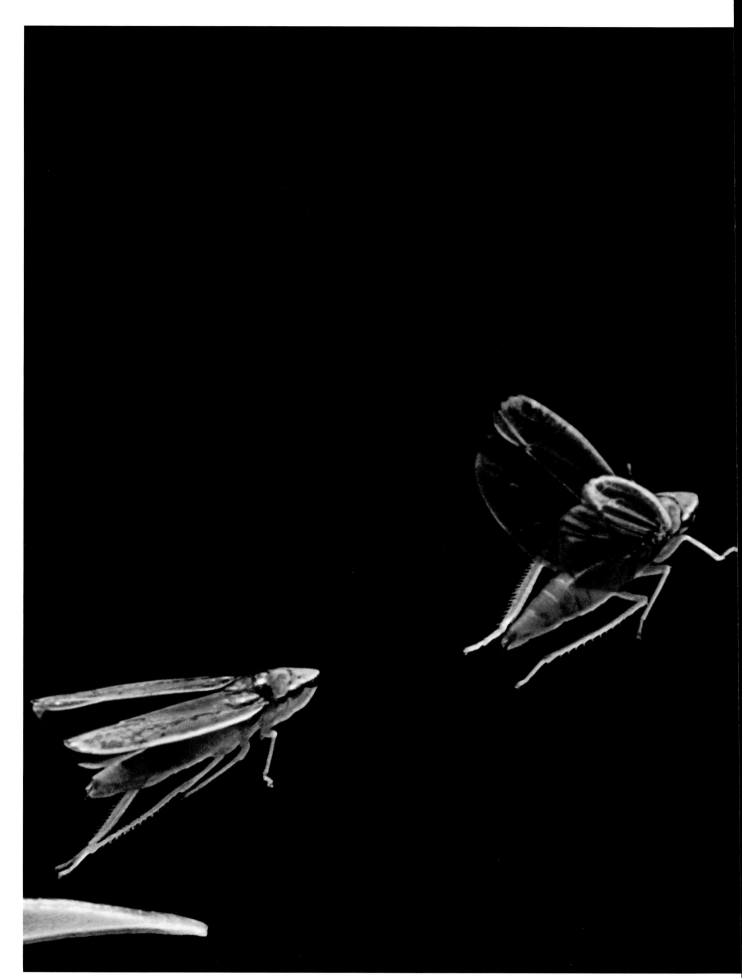

Plate 13

Plate 14

Plate 16

Plate 15

Plate 17

Plate 18

Plate 2

Plate 19

Plate 20

Plate 23

Plate 24

Plate 25

Plate 26

Plate 27

Plate 28

Plate 29

Plate 30

Plate 31

Plate 32

Plate 33

Plate 34

Plate 35

Plate 36

Plate 37

COMMENTARY ON THE PLATES

ODONATA

DRAGONFLIES AND DAMSELFLIES

Plate 1
ODONATA *Aeshnidae* Darner Dragonfly
This large dragonfly was locally common in the hammocks and pine forests of Everglades National Park, Florida. Although not very active during the day, as dusk approached these beautiful creatures began a steady patrol of the woodland rides, hunting mosquitoes. At times they would fly only a few feet off the ground, their thin chocolate-brown bodies and amber-tinted wings merging with the gloom of twilight, only the faintest rustling in the evening air betraying their presence.

In the fading light they flew higher and higher, silhouetted against the darkling sky, dodging the gigantic webs of the Golden Orb spider and weaving between the branches of the tall pines.

The exact identification of dragonflies, or, for that matter, any insect, can be a difficult and specialist task, requiring a close scrutiny of the actual specimen. As the Park authorities required me to return all insects unharmed to their favored haunts, I have been unable to name this gentle marauder with accuracy, which is aggravating to me, though of no consequence to the mosquitoes it was diligently seeking.

Plate 2
ODONATA *Sympetrum striolatum* Common Sympetrum
The Common Sympetrum abounds in England during late summer and autumn, its numbers reinforced by large swarms migrating from the Continent.

Like most dragonflies of this family it is extremely restless, returning again and again to the same spot after short and rapid flights. Sympetrum dragonflies are particularly fond of basking on roads and garden paths, and are frequently seen a long way from water.

During egg laying the male and female fly in tandem formation and dip repeatedly over some chosen pond, the female slapping the surface with the tip of her body, washing away the eggs to their underwater home. After hatching, the spiny nymphs dwell among the mud and weeds, completing their growth in one year.

I found this particular sympetrum in the attractive gardens of Sheffield Park, south of London, and spent many hours trying to persuade it to fly in the right path for my camera. Despite all my patience I was able to take only three exposures, but by luck this shot, one of three, was exactly what I sought.

Plate 3

ODONATA *Libellula flavida* Skimmer Dragonfly

Catching the glint of the sun off the waters of Florida Bay, these Everglades dragonflies roamed in bands of twelve or more along the beach, darting among the palm trees in search of insect prey. Although not the most highly colored among dragonflies, they had a restrained, soft quality I liked, and I found their habit of group flying fascinating to watch. By late afternoon they were always gone, giving way to the dusk prowlers of their order.

Plate 4

ODONATA *Perithemis seminole* Amberwing Dragonfly

This small Amberwing dragonfly, saffron tinted and very shy, took half my curiosity one afternoon in the Everglades. It was hovering and darting on lush foliage close to the water's edge, where torpid alligators, wallowing snout-deep, claimed the rest of my attention. Nearby, as if to tantalize me completely, long-necked anhingas, just returned from a fishing trip, perched high in the trees, spreading their sodden wings to dry.

It was hot and very humid, and I urged the Amberwings quite vocally, but with little avail, to disperse the milling mosquitoes feeding contentedly on my unprotected skin. It was an afternoon I remembered well into the night.

Plates 5, 6, 7

ODONATA *Lestes sponsa* Green Lestes Damselfly

This finely structured, helicopter-like flying machine, looking almost like a visitor from outer space, is a most delicate, shimmering member of *Odonata*, dancing among the luxuriant vegetation bordering ponds, ditches, and still waters throughout Europe. The metallic green bodies and dark cobalt-blue eyes of the males contrast with the duller, more bronze-green, and thicker bodies of the females. Both rarely leave the fringes of their aquatic world, the female inserting her eggs in the stems of plants well below the water's surface, where the slim green-and-brown nymphs hatch in the spring, developing among the water plants, and completing their growth in four or five months.

For ten days I tried to get this shot, slopping about in a small pond in Ashdown Forest, in the south of England, to collect a half dozen good specimens. At the end of each day's photography I returned the damselflies to their pond—not wishing to interrupt the breeding of these elegant creatures—and collected another batch for the next day's work.

I wanted to photograph the damselfly head on, right in the center between the stones, and with its wings absolutely sharp in the X position at the limit of their span. It took 252 exposures before I was satisfied with this result, which clearly shows the two pairs of wings moving quite independently.

Plate 8

ODONATA *Coenagrion puella* Common Coenagrion Damselfly

The Coenagrion damselfly hovers low over ponds and streams where it rests frequently on floating vegetation, or searches diligently for food or a mate among reeds and ferns at the water's edge. Like dragonflies, damselflies are carnivorous, though they seldom catch prey on the wing, instead picking up gnats, midges, and other small insects when they settle nearby.

This lovely blue example of Coenagrion shows clearly the enormous compound eye, usual in *Odonata*. In some species the eye may contain as many as 30,000 facets, each equipped with its own lens and retina, enabling vision almost through 360 degrees.

ORTHOPTERA

GRASSHOPPERS AND CRICKETS

Plate 9

ORTHOPTERA   *Schistocerca gregaria*   Desert Locust

Locusts are really short-horned grasshoppers which have become strongly gregarious under certain conditions. The female Desert Locust lays several batches of about seventy eggs beneath the surface of the sand, where they remain for about twenty days, depending on temperature. During this time the vegetation above the ground is also growing, so that by the time the eggs are ready to hatch there is an abundance of food for them.

On hatching, the hoppers wriggle to the surface, where the heat of the desert sun stimulates them into activity and they start feeding on leaves and grasses. During these early stages their future is determined. If there is ample living space, and they are not forced together in overcrowded environmental conditions, they develop into the solitary phase, living like grasshoppers in the area in which they were born, and not forming swarms. Locusts in this solitary phase can be distinguished by their light-green color.

But if conditions are crowded, the gregarious phase develops and the hoppers become darker in color, collecting in bands which grow larger and larger as they are joined by other unnurtured malcontents. About thirty days after hatching these hoppers molt for the last time, to become winged adults, and their swarming begins.

Swarming locusts are not only changed in color, but they are structurally different, with wings and hind legs longer than those of their solitary counterparts. Until this was determined by entomologists, the locust in each phase was regarded as a separate species.

This Desert Locust is in its darker, migratory phase, ready to join with millions of its fellows and take off in search of greener pastures.

Plate 10
ORTHOPTERA *Stagmomantis* species   Praying Mantis
Mantids are predatory insects, and their spiny forelimbs are specially adapted for seizing and holding prey. They do not pursue their victims, but lie in ambush, waiting for insects to settle within reach. Mantids have insatiable appetites and will continue catching prey even when they are full of food. They have been known to capture small birds, lizards, and frogs. To complete this list of endearing traits, the female frequently devours the male after or even during mating.

In North America the most common species were introduced; the European Mantis, *Mantis religiosa*, and the larger Chinese Mantis, *Tenodera sinensis*. Both of them are even found in the center of New York City.

Mantids, like grasshoppers, use direct flight muscles for powering their wings and do not beat them in unison. The photograph clearly shows the forewings out of phase with the hindwings.

I found this mantis on the door of my room in the Everglades. Considerate of it, for it was the only adult mantis I saw during six weeks spent in that National Park.

Plate 11
ORTHOPTERA *Melanoplus* species   Short-horned Grasshopper
Dr. Nick Jago, an expert on American grasshoppers, identified this species as belonging to the genus *Melanoplus*, a relative of the Rocky Mountain Locust, *Melanoplus spretus*. He could not be more precise than that, as many closely related species of the genus live in Florida, distinguishable only by differences in their internal genitalia!

Not very much is known about the group, but enough to have established that there are still a number of unnamed species. It is even possible that this one is new to science. Notice how it is knocking pollen from the grass in take-off.

HEMIPTERA

BUGS

Plates 12, 13, 14

HEMIPTERA   *Graphocephala coccinea*   Rhododendron Leafhopper

Leafhoppers are extremely active leaping bugs and, like grasshoppers, have long and powerful hind legs. They earned their common name, dodgers, because when alarmed they will often run around to the other side of their leaf or twig and peep out a few moments later to see if all is clear. If the situation still looks uncertain, they will retreat once more, and if things become really dangerous they will then half leap and half fly to a safer area.

Although these creatures flap their wings and fly in conventional manner when moving comparatively long distances, the multiple flash photography showed that they do not use their wings until several inches after take-off.

The Rhododendron Leafhopper is common throughout America and was not introduced into England until 1935, when it was first found in Surrey. It has since spread throughout the southern counties.

I had no difficulty in finding this specimen, for throughout the summer and autumn the one and only rhododendron bush in my garden is alive with these amusing little insects. They are reputed to spread a fungus disease, bud blast, which destroys the buds. Nonetheless, my rhododendron bush is the healthiest and most colorful in the district!

Plate 15
HEMIPTERA *Acanthocephala granulusa* Squash Bug
This insect, looking like something out of science fiction, is so named because a well-known American species attacks squashes and other gourd plants. They are also known as leaf-footed bugs, for in some species the tibiae of the hind legs are dilated and leaf-like.

There are more than 2000 species of Coreid bugs, some carnivorous, some vegetarian. Many exude a most unpleasant scent when handled.

Most of them are relatively large insects, but clumsy fliers displaying little precision when airborne. I noticed this particularly in their landings, when they seemed almost to collapse onto the leaf or bush.

I soon learned that this insect would make no effort to fly until the air temperature was at least 80 degrees F. And to obtain this photograph I had to shut off the air conditioning and spend a sticky afternoon coaxing *Acanthocephala granulusa* into the air.

Plate 50
HEMIPTERA *Miridae* Capsid Bug
The *Miridae* family of plant bugs contains some 5000 species. Most of them are fragile, soft-bodied insects, differing widely in shape, structure, and diet.

Although a few Capsid Bugs are predatory, most feed on the sap of plants, and particularly on unripe fruit—the choice of the Capsid in this photograph.

NEUROPTERA AND MECOPTERA

LACEWINGS AND SCORPION FLIES

Plates 16, 17, 18

NEUROPTERA   *Chrysopa* species   Green Lacewing

The first hint of the particular and unusual capabilities of these lovely insects in flight came to me when I was testing some equipment that had just been modified. On developing the film of a green lacewing taking off from a haw-thorn leaf, I was surprised to see that the insect had risen vertically from the leaf, with its body perpendicular and its four wings arched above, like an umbrella blown inside out.

A year later, after I had begun using multiflash, I spent many weeks trying to recreate the situation and produce three sharp images in one frame of a lacewing taking off vertically, with one in the sequence showing the inside-out-umbrella effect. The chances of achieving this were extremely slim, for lacewings do not always perform vertical take-offs, and when they do it is highly unlikely that the three images will be in sharp focus, for the creatures do not appreciate the problems of depth of field. Additionally, the chance that the flashes will fire at the precise moments the insect adopts these fascinating positions is even more remote.

In all, I took about 900 exposures until finally, when my patience had very nearly come to an end, I more or less achieved what I was after, obtaining a number of intriguing multiflash pictures that show how lacewings fly. As well as performing vertical take-offs (Plate 16), they frequently do loops or half loops (Plate 17), often landing directly behind the position in which they were originally, or occasionally landing on the same spot.

I wondered whether such antics are intended to baffle pursuers, or whether they are the result of pure chance due to lack of flight control, or poor navigation. And I came to the conclusion that they are controlled, purposeful maneuvers.

Plates 19, 20
MECOPTERA *Panorpa* Scorpion Fly
The frontal view of the scorpion fly looking very preoccupied, on Plate 19, shows the wings returning upward after a downstroke. In the fully down position, the hindwings are folded so far forward that their tips actually touch. On Plate 19 the insect is caught a split second after its hind feet have left its perch.

I took three separate shots of this scorpion fly taking off, and in each one its legs were crossed at this moment in flight. Was it an elegant entrechat for my delight, a quirk of this particular scorpion fly, or the usual deportment of the species?

LEPIDOPTERA

BUTTERFLIES AND MOTHS

Plate 21

LEPIDOPTERA *Gonepteryx rhamni* Brimstone

The very name "butterfly" originated from this lovely sulfur-yellow insect, in medieval days called the "butter-colored-fly." Brimstones are long-lived insects, those that emerge from the chrysalis early in August sometimes surviving until the following June. Winter is spent in hibernation, the butterflies hiding in thick clumps of ivy or other evergreen bushes where their leaflike wings blend with the surroundings. In England the Brimstone is one of the first insects to emerge in the spring, and it can often be seen flying on warm February days before the last remnants of snow have disappeared from the northward slopes.

Although really a woodland butterfly, the Brimstone also favors the open countryside, where the males can be seen chasing the females along the hedgerows, often traveling miles in quest of a mate. In due course the female deposits her eggs on the young leaves at the tips of buckthorn shoots.

In flight, the pale greenish-white female can be mistaken for the "cabbage white," its close relative, but it has a stronger and more direct manner of flying. The photograph shows a male, with its wings at the limit of an up-stroke, flying over the spring flowers of the forsythia.

Plate 22
LEPIDOPTERA   *Aglais urticae*   Small Tortoiseshell
The Small Tortoiseshell is a member of the largest and most widely distributed family of butterflies in the world, the *Nymphalidae*, or brush-footed butterflies. From the north temperate zone it includes such familiar and boldly colored insects as the Fritillaries, the Comma, (Anglewings), the Painted Lady, the Peacock, the Camberwell Beauty (Mourning Cloak), and the Admirals.

The Small Tortoiseshell is one of the commonest butterflies found in England and Europe, its territory stretching through Asia as far as Japan. It seems equally at home in city gardens, woodlands, alpine meadows, or on the banks of mountain streams. Like the Brimstone, it hibernates, though preferring barns, attics, cellars, or hollow trees.

The Small Tortoiseshell courtship ritual in spring often takes place on the ground or on a warm wall, the butterflies facing each other, the male stroking the wings and antennae of the female with his own feelers. The caresses are interspersed with short flights, after which the pair repeats the performance on the same spot.

This photograph shows the dark undersides of the wings as they approach the end of their backward stroke.

Plate 23
LEPIDOPTERA   *Vanessa atalanta*   Red Admiral
This well-known butterfly is international in its distribution, for it is found not only on both sides of the Atlantic but also in New Zealand and Haiti. In the Old World it ranges from North Africa right across Europe and into Asia as far as Iran, and in the New World from central Canada through the southern U.S.A. to the Antilles and Guatemala. The Red Admiral is unable to survive the winter in countries with a cold climate, and it recolonizes from the south each spring. It is a fast and powerful flier, and in the autumn many which have bred in England make their way south again toward the Mediterranean, often flying at night.

The caterpillar feeds on nettles and related plants, making leafy tents spun together with silk in which they pupate when fully grown.

The Red Admiral likes flowers, particularly Michaelmas daisies and buddleia. I found this one in my garden during late summer feeding on the nectar from a white buddleia.

Plates 24, 25

LEPIDOPTERA *Papilio machaon* European Swallowtail

The swallowtail family, *Papilionidae*, contains some of the largest and most spectacular of all butterflies. Among them are the birdwings from the forests of southeast Asia and Indonesia, with a beauty of wing and flight so breathtaking that those who see these creatures in their natural setting never forget the experience.

The European Swallowtail, *Papilio machaon*, is in many countries the only native representative of its family, and one of the butterflies best known to laymen and naturalists alike. This species occurs throughout Europe, North Africa, Asia, and North America, while in Tibet it can be found near Mount Everest at altitudes above 15,000 feet.

The strong flapping flight of the Swallowtail is seen all too rarely in England, where it is in danger of extinction. I have selected these two photographs, not only to display the full beauty of this fine butterfly, but also to give some idea of its wing power in flight.

Plates 26, 27

LEPIDOPTERA *Danaus plexippus* Monarch

The Monarch belongs to the milkweed group of butterflies, (*Danaidae*). All the butterflies in this family are relatively large and, as many are poisonous, they are seldom attacked by predators, flaunting their inedibility in their colorful and boldly patterned wings and bodies.

Although slow, the Monarch is powerful on the wing, with a characteristic soaring, gliding flight. It has a reputation for long migratory journeys. During autumn in the United States swarms of Monarchs can be seen making their way southward, often touching down to rest in large cities.

Monarchs spend the winter in tropical or subtropical areas, and those that survive drift northward again in the spring, laying eggs as they go. The Monarch is found throughout the world, though nobody is certain to what extent this may be due to man's help. Monarchs regularly turn up in Western Europe and Africa, and have even spread across the Pacific to Australia. Specimens have been seen flying hundreds of miles out to sea, so at least a few cross the oceans under their own power.

What a spectacular butterfly this is. I photographed it in Florida and was entranced by its superb form and beauty.

Plate 28

LEPIDOPTERA *Calpodes ethlius* Brazilian Skipper

There has always been some doubt as to whether Skippers (Hesperiidae) are, in fact, butterflies, for they share a number of characteristics with moths. They certainly look mothlike with their short hairy bodies and prominent eyes. Some fold their wings flat over their bodies like moths, and one species found in Australia even has a frenulum, the wing-coupling device characteristic of the moths. They also build primitive cocoons, as do moths.

To discount all this, the knobbed antenna is a true butterfly trait, and Skippers are day fliers. Although the majority are small and somewhat dull in color, there are a few species, from the American tropics, which are brilliantly iridescent. Unlike most butterflies, Skippers, particularly the large exotic species, are capable of dazzling performance on the wing. They are the supersonic jets of the butterfly world, capable of very fast acceleration and sustained high speed, and their thick and streamlined bodies house large, powerful wing muscles.

The Brazilian Skipper is one of the larger Skippers found in the United States, where it is common in the Deep South. The pale markings on the wings are transparent spots. I saw this insect flying toward me over some mangroves at a speed I would never have thought possible. It flew straight past, after twenty-five yards or so made a sudden about-face, and with a series of rapid maneuvers landed on a flower at my feet, where it calmly fed on the nectar.

My first reaction was that there was not much point in photographing it, as even my special rapid shutter would not cope with the speed. Also I did not have a net with me, so I had to catch the Skipper between my thumb and forefinger and put it in a matchbox! I was very pleased to get this shot, for, as luck would have it, I did not see another Brazilian Skipper during my entire stay in Everglades National Park.

Plate 29
LEPIDOPTERA *Danaus gilippus berenice* Queen
Like the Monarch, the Queen is a member of the milkweed family of butterflies and is easily distinguished from the Monarch by its darker, more crimson ground color. The Queen is commonly found only in the southeastern United States. It has a distinctively slow, sailing manner of flight, quite apparent in this shot.

Plate 30
LEPIDOPTERA *Agraulis vanillae* Gulf Fritillary
The coloration of this Heliconian butterfly is very distinctive, with its ground of bright orange on the upper wings, and its underwings brilliantly marked with oval-shaped metallic silver blotches.

It is a fast flier and like so many butterflies is addicted to flowers. It is sometimes found as far north as New Jersey, but is generally rare in the northern United States.

Plates 31, 32
LEPIDOPTERA   *Heliconius charitonius*   Zebra

The Zebras belong to the Heliconian butterflies, a truly tropical family characterized by brilliant coloring and bold patterns. They possess poisonous and distasteful body juices which, together with their bright wing colors, make them easily avoided by their enemies. Other species of butterflies, not only from the *Heliconii* but often from entirely different families, have taken advantage of this warning coloration and have adopted similar bold patterns to gain protection.

Only three members of the *Heliconii* occur in the United States, and these are confined to the South. The Zebra cannot be confused with any other butterfly. Its flight is slow and fluttering, with a very shallow wingbeat. It can glide from the tops of trees and usually flies in the company of others of its species. In the Everglades it was a beautiful sight to see a group dancing through the sunlit mangrove swamps.

Plate 33
LEPIDOPTERA   *Leucania comma*   Shoulder-striped Wainscot

In this multiflash exposure of a Shoulder-striped Wainscot taking off, a shower of scales can be seen falling from the insect. At first I thought the spots were an imperfection on the transparency, but on closer inspection the origin of the spots became evident, for I was able to determine the shapes of the individual scales.

Scale shedding is not unknown in the insect world. Bee Hawk moths, for example, lose all the scales from their wings during their first flight, leaving the wings transparent, except for a thin border around the edge.

The Shoulder-striped Wainscot is a typically dull and inconspicuous Owlet moth which rests during the day among dead grasses and reeds, but this one at least distinguished itself by revealing to me an unknown phenomenon.

Plates 34, 35
LEPIDOPTERA   *Noctua pronuba*   Yellow Underwing

In this take-off shot the antennae of the Yellow Underwing are bent back by the airflow. Again, a cloud of scales is falling from the wings as the insect becomes airborne. There is a 1/12 second interval between the three different wing positions.

This moth, like the Broad-bordered Yellow Underwing in Plate 36, comes from the largest family of moths, the Owlets, and although some tropical species are brightly colored, with wingspans of ten or twelve inches, the large majority are drab, medium-sized moths with cryptic wing patterns that merge into natural surroundings.

Many species of Owlets make long migratory journeys; the Silvery Owlet *Laphygma exigua*, for example, flies from North Africa to breed in England.

Plate 36
LEPIDOPTERA *Lampra fimbriata* Broad-bordered Yellow Underwing
The Broad-bordered Yellow Underwing is a member of one of the most striking groups of Owlets and, like all Underwings, has an ingenious way of baffling its enemies. During the daytime it rests on the ground or the bark of a tree, where the colors of its forewings make it inconspicuous. The concealed hindwings, however, are boldly colored with rich orange-yellow and jet-black bands, and, when disturbed, this inconspicuous moth suddenly transforms into a flash of brilliance and flies off in an erratic manner, confusing its enemy. At a safe distance it settles again, closing its wings and virtually disappearing once more from view.

In this shot the flight control of the wings is beautifully evident as the moth changes course.

Plate 37
LEPIDOPTERA Unidentified species Owlet
In addition to their scale dropping, another aspect of insect behavior became particularly evident in the case of the smaller moths. My camera showed that many of them frequently fly on their sides or even completely upside down, but it would be necessary to take a movie sequence to establish how long moths maintain such attitudes. At this particular moment, the noctuid appears to be performing a barrel roll.

Possibly such "abnormal" flight occurs either when an insect evades some obstacle or takes off from an upside-down surface. Most larger animals maintain balance because the force of gravity acts on special balancing organs—located in the inner ear in mammals and birds. But gravitational forces have far less effect on small insects, and it even seems likely that they do not use gravity as a reference for maintaining a horizontal flight position, relying instead on other environmental stimuli or on some sense not yet determined by entomologists.

Plates 38, 39

LEPIDOPTERA   *Celerio euphorbiae*   Spurge Hawk (Sphinx)

The Spurge Hawk is one of southern Europe's loveliest moths, the upper surface of its wings marked with various shades of green, brown, pink, and black. I once saw this insect in the dusk of a summer's evening on a Mediterranean beach. It was flying with surprising gentleness and quiet, hovering on quivering wings as it sipped the nectar from oleander flowers. Moving from one blossom to another, it then quite suddenly flew out to sea in the gathering dark.

Hawk moths have never failed to excite me. Their front wings are long, narrow, and pointed, their hindwings small. They have thick and beautifully streamlined bodies, packed with powerful flight muscles. They are the falcons of the insect world. Hawk moths are sometimes called hornworms (the caterpillar of most species has a horn on its back), but this name is an insult to such a splendid creature.

These moths have very long tongues and are able to reach the nectar of tubular flowers. Certain flowers have developed such a special relationship with some Hawk moths that they rely on them completely. In 1891 Alfred Russell Wallace found an orchid in Madagascar that had its nectaries ten or eleven inches below the entrance of the tubular flower. He was puzzled as to how the flower was being cross-pollinated, there being no known insect with a tongue long enough to reach the nectary. Wallace predicted that a Hawk moth with a sufficiently long tongue must exist on the island, and twelve years later, in 1903, the moth was discovered. Appropriately enough, it was named *Macrosilia morgani predicta*.

As against this, one species, the Death's Head Hawk *(Acherontia atropos)*, has an unusually short tongue, and instead of visiting flowers, it manages to enter the nests of bees, pierces the honeycomb, and sucks the honey.

The photographs show the moth shortly after take-off, one with its wings in the near vertical position. Insects often rotate their wings in this way when they fly slowly or hover. The second picture shows the full beauty and power of this magnificent moth.

Plate 40

LEPIDOPTERA *Lithosia lurideola* Common Footman

The Common Footman belongs to the *Arctiidae* family of moths, whose members show considerable variation both in appearance and size among the species. They are called Footmen because when at rest the moths wrap their long wings tightly around their bodies—like a footman's cloak.

The Footmen are flower visitors, with well-developed tongues. This one is homing enthusiastically on a succulent yellow blossom.

Moths of this family are not only capable of detecting the ultrasonic squeaks of bats, but they also have the almost unbelievable ability to emit ultrasonic sounds of their own, which confuse the bats' radar detection system. An extremely sophisticated evolutionary refinement.

Plate 41

LEPIDOPTERA *Saturnia pavonia* Emperor Moth

The Emperor moth is one of the few representatives of the Giant Silk moth family (*Saturniidae*) to occur in Europe. Most Silk moths are huge, exotic-looking insects, with prominent crescents or eyespots on their wings. Among the largest species are the Atlas and Hercules moths with a wingspan of ten inches or more, while in North America, the home of some sixty species, the largest is the Cecropia moth *(Hyalophora cecropia)*. In comparison, the commercial silkworm moth of China, which belongs to a different family, is a dull, uninteresting-looking insect, now unable to fly as a result of selective breeding.

The Emperor is the only Silk moth which breeds in England and, as the caterpillars feed on heather, it is found most often in open country on heaths and moors. The male flies by day, equipped with large, feathery antennae that enable him to pick up the scent of virgin females several miles distant. To watch this is an almost uncanny experience. I have placed a newly emerged female Emperor in a muslin cage and, within minutes, have seen up to a dozen males appear in the distance and fly straight toward the caged female.

Plate 42

LEPIDOPTERA *Gonodontis bidentata* Scalloped Hazel

The Scalloped Hazel is a member of the second largest family of moths, the Geometers. In America they are generally known as measuring worms, and in England looper moths. These names are appropriate, for the caterpillars are unusual in having feet only at the front and rear of their bodies, obliging them to progress with a looping gait, which makes them appear to measure the ground they cover.

Many moths at rest look much like shriveled leaves, and the Scalloped Hazel adds to this resemblance with the scalloped edges to its wings. The species varies considerably in its ground coloring, from light brown to nearly black. This one, of a pale and slightly yellowy gray, shows itself off beautifully as it veers away from the foliage.

Plate 43
LEPIDOPTERA *Deuteronomos alniaria* Canary-shouldered Thorn
The Canary-shouldered Thorn moth flies in the autumn over wooded areas and during the day rests on fences or tree trunks. This cheerful little fellow is a male, identifiable by the mildly plumed antennae, which fired the flash for this shot.

He has the relatively broad wings usual among the Geometers which, although quite fast in flight, lack the control and sustained power of many of their relations.

Plate 44
LEPIDOPTERA *Phlogophora meticulosa* Angle Shades
One of my favorite Owlet moths is the Angle Shades, but unlike most insects its subtle beauty can be fully appreciated only when the creature is not flying. At rest, its wings, washed with delicate shades of green and pink, wrinkle up like folded fabric.

The moth is found during most months of the year, and in Britain its numbers are increased by migrants from across the English Channel.

Plate 45
LEPIDOPTERA *Platyptilia pallidactyla* Plume
Technically, the Plume moths belong to the *Micro lepidoptera*, which are simply small moths, such as the Clothes moth.

Plume moths have very narrow feather-like wings, deeply divided into a number of sections, and long, fragile, spider-like spurred legs. These delicate little insects have a pathetically weak and slow flight, and how or why they evolved so apparently inefficiently might seem to be a mystery. But somehow Nature has found a niche for them, for there are several hundred species and they occur all over the world, even in the bleak 15,000-foot environment of the Andes.

Plate 46

LEPIDOPTERA   *Composia fidelissima*   Pericopid

Although this is technically a day-flying moth, I found it early one evening in the Everglades and was not aware of its sensational colors until I took it into the light of my room. Like so many tropical insects, this moth gains protection from enemies by displaying warning coloration. When caught or disturbed, it further baffles its foes by shamming death, folding its wings backward and discharging a yellow, foamy liquid over its body, at the same time making a bubbling noise. To what limits of deceit will a beautiful moth go! This Pericopid is found in Cuba, Florida, and Brazil.

COLEOPTERA

BEETLES

Plate 47

COLEOPTERA   *Strangalia maculata*   Spotted Longhorn

In comparison with the Cockchafers, the Longicorns or Longhorn beetles are slender, elegantly shaped insects, all legs and antennae, which together are often several times the length of the body. The Longhorns also show an infinite variety of color and design and vary greatly in size, the smallest being about a quarter of an inch long. The largest, from South America (*Titanus gigantens*), has a body length of six inches, and its antennae extend its total length to twelve inches.

The family includes striking mimics, such as the black and yellow Wasp beetle, which not only looks like a wasp but even moves like one, with hurried, jerky movements so characteristic of wasps. Many species of Longhorns have a peculiar habit of sitting quite motionless in a cataleptic state. However, if they are picked up, they can surprise you by not only making loud creaking noises but also biting hard with their powerful jaws.

Beetles are frustrating insects to photograph in flight, as most species are extremely reluctant to take to their wings, and even when they do become airborne, they lack directional stability. Longhorns, however, are strong fliers and readily fly in warm sunshine. I found this one feeding on the nectar of a wild rose in my garden, and I persuaded it to fly in captivity by exposing it to the hot afternoon sunlight slanting through the window of my studio.

Plate 48

COLEOPTERA *Melolontha melolontha* Cockchafer

The Cockchafer, Maybug, or June beetle, as it is variously called, is a member of the Scarab family, which includes some of the world's largest and most impressive beetles. The giant African Goliath beetle and the five-inch-long Hercules beetle from South America are among them. The Scarab beetle, so sacred a symbol to the ancient Egyptians, is also, of course, a member of this family.

Cockchafers are heavily built beetles with short antennae adorned with clubs formed by thin movable plates which they can spread out fanwise. The adults of many species, in addition to feeding on nectar, do considerable damage to flowers and foliage when present in large numbers. The dark bronzy-green Japanese beetle, imported into North America in 1917, is notorious for this behavior.

The Cockchafer cannot fly until it has filled numerous air sacs in its body. Even then it has difficulty in becoming airborne and is a slow, noisy, clumsy flier. Particularly attracted to lights on warm evenings during early summer, it bangs against windows or, entering and circling a room, often blunders into objects and falls to the floor.

Looking almost like some furry little animal in flight, this Maybug performed faultlessly in front of the camera.

Plate 49

COLEOPTERA *Cantharis pellucida* Wood Sailor

Unlike the majority of beetles, the Soldier and Sailor beetles (*Cantharidae*) have surprisingly soft bodies—even their wing cases, instead of being tough and hard, are more like damp parchment paper. They also have narrow bodies and longish legs. The Wood Sailor, together with many similar species, can be seen during the day busily walking about on plants or delving into blossoms and feeding on pollen or nectar.

In this photograph the Wood Sailor is taking off from a plantain flower. The wingbeat frequency was measured as being fifty-two cycles per second. In the first image the wings are seen slicing downward through the air, while in the second image, 1/30 second later, the wings have completed a further one-and-three-quarter cycle and are close to the limit of their upstroke. The third image shows the wings bent 90 degrees downward beneath the body, clearly demonstrating the exceptionally large stroke angle of beetles. Notice that the elytra are oscillating through about 30 degrees in synchrony with the wings. The first image also shows the hind legs underneath the body as the beetle kicks off, followed by two images in which all six legs have been retracted.

Plate 50
See Page 89

Plate 51
COLEOPTERA *Coccinellidae* Ladybug
The Ladybug, Ladybird, or Ladybeetle, as it is variously called, is a really true friend of man, for it preys quite ruthlessly on the aphids, greenflies, and scale insects that threaten our crops. There are more than four thousand species of Ladybug, and, with very few exceptions, they work on our behalf in both the larval and adult stages. That they are considered lucky is no poetic euphemism—they are lucky!

In legend they have a somewhat disturbed home situation, so be especially kind to them, for they work for us in a way that even the most powerful insecticides cannot match. Spraying aphids must have a percentage of failure; eating them has none.

The ladybird in this photograph, apparently dancing the light fantastic on the stamen of some preferred flower, is in fact about to fly away, leaving behind a shower of yellow pollen as she begins her homeward flight.

HYMENOPTERA

BEES, WASPS, AND ANTS

Plate 52
HYMENOPTERA *Ophion luteus* Yellow Ophion Ichneumon Wasp
The Yellow Ophion, together with a bewildering collection of other Ichneumon, Braconid, and Chalcid wasps, are nearly all parasitic on insects and spiders. The female wasp tirelessly searches for a specific host and, on finding it, pierces it with her ovipositor and lays an egg (or eggs, depending on species) on or within it. In due course the larva hatches and starts to feed on the tissue of its host, avoiding all vital organs, thus enabling its victim to grow to full size. Eventually it spins a cocoon and pupates within, to emerge later as an adult wasp.

The Yellow Ophion Ichneumon is familiar to many. Attracted by light, it often flies into houses at dusk. It is orange in color and about three-quarters of an inch in length. Unlike many Ichneumons, the female has no long ovipositor.

Not entirely likable, this insect, but it leads its own life most efficiently, however bizarre that life may seem to us.

Plate 53
HYMENOPTERA *Sceliphron caementarium* Mud Dauber Wasp
Unlike the social bees and wasps, the Mud Dauber leads a life on its own, and as in the case of most solitary wasps, the female, after making her nest, hunts for spiders. Solitary wasps feed their larvae on whole spiders, which they first paralyze by stinging. The injected venom has antiseptic properties, so that the prey remains fresh for several days or even weeks.

The wasp carries the helpless victim back to her nest and, after laying an egg, entombs both the victim and the egg by sealing the cells. When the egg hatches, the larva has a fresh supply of protein on which to nourish itself over the following days, until it reaches maturity.

The Mud Dauber is noted for the long thin stalk of its abdomen and for its long hind legs. The photograph shows this extraordinary aerial complex on the hunt for likely victims. That its small wings can keep so separated an insect airborne seems in itself a remarkable aerodynamic feat.

Plates 54, 55, 56, 57
HYMENOPTERA *Polistes metricus* Paper Wasp
Paper wasps are familiar to country dwellers in North America, often building their open-celled nests under the eaves of houses and barns. Unlike Honeybees and Common wasps, which have only one queen, many species of Paper wasps tolerate several queens in one colony.

Despite their ominous appearance, most Paper wasps are good-tempered and sting only when severely provoked.

Plate 58
HYMENOPTERA *Bombus agrorum* Carder Bumblebee
Bumblebees differ from Honeybees in that their colonies do not survive the winter. Only the fertilized queens live through to the following year to start a fresh community.

In the spring, unaided by workers, each queen selects a suitable hole, usually underground, and makes a nest of fine grass, moss, and leaves. She then prepares a cake of honey and pollen on which she builds a circular wall of wax, laying a batch of eggs inside it and covering them with a lid. She also fashions a waxen pot about the size of a thimble, which she fills with honey before settling down to brood the eggs. The eggs hatch in about four days, and the young larvae eat the food provided. After three weeks the first workers hatch and begin to assist the queen. The community enlarges quickly, and both males and young queens are reared later in the year to perpetuate the species. The old queen, her workers, and males all die at the end of the season.

Like many beetles, Bumblebees tend to be clumsy on the wing, particularly when flying slowly immediately after take-off, often bumping into objects and stalling. Yet, in spite of this apparent weakness, they can frequently be seen flying on cold, wet, and windy days, visiting one blossom after another with dogged determination, long after all other insects have been grounded.

This worker is very near the end of its life, its tattered wings frayed from constant use.

Plates 59, 60
HYMENOPTERA *Apis mellifera* Honeybee
There are three different species of Honeybee, and, of the three, *Apis mellifera*
has been successfully exploited by man for thousands of years. Honeybees
live in organized communities of about 80,000 members. In a colony, which
consists mainly of workers—that is, sterile females—only the queen is capable of reproduction, while the purpose of the male, or drone, is to fertilize
the queen.

The workers do all the work in the hive. They feed the larvae, the drone,
and the queen, build the wax cells, clean the hive, and fight off intruders. Only
the older and more experienced workers collect the pollen and nectar.

In summer a worker bee lives little more than six weeks, and although bees
forage only during daylight and in reasonably good weather, activity in the
hive never ceases.

Plate 60 shows a worker bee one sunny morning, laden with yellow pollen
collected from a bank of March crocuses in my garden. The bee in Plate 59
had been gathering nectar from rosebay willow herb during a heat wave in
late summer.

Plates 61, 62
HYMENOPTERA *Vespa vulgaris* Common Wasp (Yellow Jacket)
The Common wasp is the most widespread species of social wasp in Europe.
It also occurs in North America and has become established in New Zealand.
Like the Bumblebee, its colony exists for only one season. The adults feed on
nectar and fruit juices in the same way as bees, but they have shorter tongues
and so can suck nectar only from shallow flowers. Wasps have no pollen-
collecting organs. They usually make their papery nest in a hole underground,
although sometimes they build in a hollow trunk, attic, or even a thick bush.

Plate 71
HYMENOPTERA *Tenthredo scrophulariae* Sawfly
The female Sawfly, easily distinguished from a bee or wasp by the absence of
a narrow waist between the thorax and abdomen, has her own distinctive
feature. She is equipped with a saw, comprising a pair of serrated blades, with
which she makes incisions in leaves, stems, or buds, to house her eggs. After
hatching, the larvae, resembling the caterpillars of butterflies or moths, feed
on the plant tissues.

Sawflies enjoy a brief life as adults, dying quite soon after they have mated,
yet despite this abrupt habit of demise they have successfully descended from
Triassic ancestors of some 230 million years ago.

DIPTERA

TRUE FLIES

Plates 63, 64
DIPTERA   *Scatophaga stercoraria*   Yellow Dungfly
The hairy greenish-yellow or ginger long-legged Dungflies are familiar to anyone who walks through meadows and fields, where their larvae feed on animal droppings. Although the adult visits flowers for nectar, it has the malevolent habit of attacking other insects, sometimes even its own kind, cutting the nerve cord of their necks with the sharp teeth at the base of its proboscis.

Plate 63 shows the fly using its back two pairs of legs for springing into the air during the initial stages of take-off, vibrating the leaf as it goes. In the third image the insect is climbing practically vertically, showing that it has rotated through nearly 90 degrees within 1/50 second.

In Plate 64 it is in full flight, tracking down some unwitting quarry that dares to feed on its favorite nectar.

Plate 65
DIPTERA   *Lucilia* species   Greenbottle Fly
There are a number of different species of flies which have bodies brightly colored in shades of metallic blue, green, or bronze, but their exact identification can be difficult to establish by the nonexpert. While the larvae of the majority of them are scavengers and carrion feeders, some species are parasitic.

The Common Greenbottle fly (*Lucilia caesar*) feeds on the nectar of a wide variety of blossoms in gardens and woodlands. It is particularly fond of umbelliferous plants and can be seen flying with almost unbelievable rapidity from one blossom to the next.

Plates 66, 67

DIPTERA   *Eristalis tenax*   Dronefly

Of all insect mimics the Dronefly is perhaps the most frequently seen, for it is abundant in most parts of the world. It mimics the Honeybee, and not only looks like one in color and appearance but feeds on flowers alongside the true bees.

The larva of the Dronefly is popularly known as the rat-tailed maggot; it has a long telescopic air tube that allows the creature to breathe while submerged. It lives in stagnant water or semiliquid mud in such places as gutters or neglected water troughs, where it feeds on minute particles of decaying organic matter.

In Plate 66 the Dronefly is happily emulating the Honeybee, while in Plate 67 it looks much more like the true fly it really is, at its moment of truth, a split second before hitting a spider's web.

Plate 68

DIPTERA   *Echinomyia grossa*   Parasitic Fly

Along with parasitic wasps, Tachinid or Parasitic flies are among the world's most important natural checks on a wide variety of agricultural pests. They also play a useful part in the cross-pollination of plants.

As larvae, Parasitic flies live within the bodies of all sorts of insects, including moths, beetles, and other flies. This Parasitic fly prefers large moths such as the Oak Egger. It is a particularly large fly, looking, flying, and sounding like a Bumblebee. Its favorite haunts are sandy heaths and open forest, where the adults are attracted by flowers, particularly ragwort, angelica, and wild parsnip.

This one is taking off backwards and sideways, in a nonchalant pose, from a water mint flower. Notice that to achieve this extraordinary maneuver the leading edges of the wings have twisted through almost 180 degrees, so that the trailing edges are pointing forward.

Plate 69

DIPTERA   *Musca domestica*   Housefly

There are several species of so-called Houseflies, but *Musca domestica* is the most widely distributed. Although the insect can spread disease, its larvae are useful general scavengers, for they feed on practically any decaying matter, from manure and carrion to garbage and damp newspapers. This catholic taste has enabled the Housefly to follow man wherever he goes.

The true Housefly can be identified very easily by its manner of flight. It has a characteristic way of flying contentedly around a room for long periods, often beneath lights. This flight pattern is made up of a number of irregular triangular or quadrilateral courses, with an almost imperceptible hovering taking place at the curves, while the sides are covered in a rapid dart. When alone and undisturbed, these flies maintain a more or less constant height and regular course, but when several decide to patrol the same beat, one will often rapidly dart toward another. After a short flurry patrolling begins again.

This photograph took a number of weeks during winter to obtain. It was early on in the project when I was experimenting on shutters and triggering systems. At this time of the year houseflies are difficult to find, so I had to obtain them from a local pest research laboratory. I soon found that these laboratory-bred houseflies were reluctant to fly at all, as their wing muscles had had little exercise, due to the unnatural and overcrowded conditions in which they had been living. They had a frustrating habit of running and jumping instead of flying. The only answer was to release them in the kitchen for a few days to enable them to become suitably agile for flight photography.

Friends who visited the house looked horrified when they saw the large number of flies I was accommodating, and my observation that the flies were probably cleaner and less contaminated than my visitors was not readily accepted. In this shot the fly is rising from a loaf of homemade bread.

Plate 70
DIPTERA *Syrphus lunigar* Hoverfly
People become apprehensive, thinking it is a wasp, when a yellow-banded Hoverfly approaches too close. In fact, these friendly and gentle little insects are of enormous benefit to mankind, sometimes saving crops from total annihilation. The larvae of Hoverflies in this group are without eyes or legs, and yet are able to wander about plants and catch as many as fifty aphids a day each. Thanks to their insatiable appetites and the reproductive powers of the adults, relatively few Hoverflies can exterminate an infestation of countless millions of aphids.

In this photograph of a typical and common Hoverfly, the three flashes were fired 1/20 second after each other, recording a mere 1/10 second in the life of the fly, shortly after take-off. During this time the insect had completed eleven and a half cycles of wing movement, at a wingbeat frequency of 115 cycles per second. In the first two images the wings are moving upward, but the final image shows the fly making an ascending turn to starboard, while the wings have completed about a quarter of their downward power stroke.

Nobody can doubt the remarkable flying ability of these insects, if one considers the ease with which they remain hovering in one spot, apparently motionless, though they are really flying in a complex of fluctuating wind currents. They achieve this apparent suspension by visually locking themselves on to a nearby landmark and maintaining a constant distance and bearing from it by instantaneous and very precise adjustments to their wings.

Plate 71
See Page 107

Plate 72
DIPTERA *Chironomous* species   Nonbiting Midge
Nonbiting Midges are paler, more fragile in appearance than mosquitoes, and have a humpback look about them. Their inability to suck blood further distinguishes them from mosquitoes and the biting midges.

The males can often be seen in swarms of thousands, even millions, in the warm evening air, dancing invitingly for the females, which fly into their midst and mate on the wing.

The larvae of Nonbiting-Midges play an essential role in the ecology of ponds and streams, consuming quantities of organic matter before becoming the main diet of many species of small fish and carnivorous aquatic insects, which in their turn form the major food supply for larger fish.

The larvae of some midges have become extraordinarily specialized in their life-style, one species being able to survive only under the wing pads of the mayfly larva, another preferring a saltwater existence 120 feet below the surface of the Pacific Ocean.

Plate 73
DIPTERA *Aedes*   Yellow Fever Mosquito
This mosquito is memorable not only in its own right but for the manner in which it was photographed.

Of all the disease-spreading insects, mosquitoes are the most notorious. They are found virtually everywhere in the world, from the Arctic to the tropical salt marshes and swamps, and may spread not only malaria but an alarming assortment of other diseases, such as encephalitis, dengue, yellow fever, and filaraiasis.

Yet, in spite of their low reputation, mosquitoes occupy a vital niche. In the Everglades, for example, the whole cycle of life, from the fish and birds to the alligator, revolves around this infuriating little fly.

To obtain this photograph, some forty specimens were released into the Everglades bedroom where I had set up my equipment. The night was not comfortable, though this shot very nearly justified the irritation that ensued.

Plate 74

DIPTERA *Physocephala nigra* Thick-headed Fly

Although this odd-looking fly is not a true Hoverfly, it is a close relation. It can be easily distinguished from Hoverflies by its long, conspicuously clubbed antennae. Those of Hoverflies are usually very short and bristle-like. Thick-headed flies (*Conopidae*) derive their name from the large blunt head which is often wider than the rest of the body. Like many Hoverflies, they mimic bees and wasps, this one resembling a Sand wasp.

As larvae, all members of the family are parasitic, on bees, wasps, and in some species locusts. The females are said actually to lay eggs on their hosts while in flight.

In this shot one haltere is clearly visible as a bright yellow blob immediately behind the thorax.

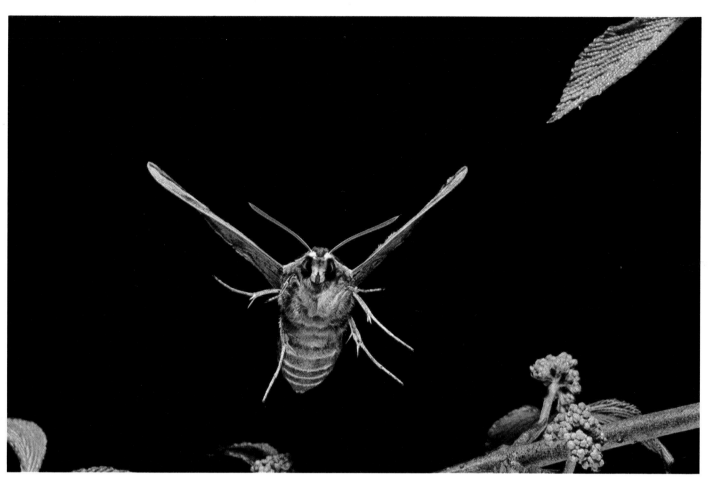

Plate 38

Plate 39

Plate 40

Plate 41

Plate 42

Plate 43

Plate 44

Plate 45

Plate 46

Plate 47

Plate 48

Plate 49

Plate 50

Plate 51

Plate 52

Plate 53

Plate 54

Plate 55

Plate 56

Plate 57

Plate 58

Plate 59

Plate 60

Plate 61

Plate 62

Plate 63

Plate 64

Plate 65

Plate 66

Plate 67

Plate 68

Plate 70

Plate 69

Plate 71

Plate 72

Plate 73

Plate 74

THE PHOTOGRAPHER AT WORK

When friends asked why I should want to photograph insects in flight, the question itself underlined the general unawareness of a small world of great beauty, which the camera lens has now discovered. While no one would query the aesthetic or scientific value of photographing birds in the air, to do the same with a fly seemed to most people an unnecessary pursuit.

Photographically, of course, the answer was that insects had never before been photographed successfully in free flight, which made it an Everest I wanted to conquer. But more than that, there was the absolute fascination I felt for the subject, so I had a dual reason that was irresistible to me, if inexplicable to others.

For as long as I can remember I have been fascinated by all aspects of insect life, and for the past twelve years I have been photographing insects at rest or crawling. Yet, for me, the most wonderful thing about insects is that they fly. Birds fly too, but insects are far more mysterious. You can quite clearly watch a bird fly; with large birds you can even follow each wingbeat and see the shape of the wings changing during different phases of movement. But you can't do that with an insect in flight. With insects, even the wings of the slowest fliers are blurred, while the paths of some insects through the air are so rapid and fluctuating that it is impossible to keep track of any image at all.

A bird in flight is a lovely, graceful creature, but insects, in their own way, are equally beautiful. And for a photographer they have even more to offer, with their enormous number of species and much greater diversity of structure.

In the main, a bird is a bird. Not quite so for insects, which have an almost limitless range of wing and body shapes, and flight mechanisms. As a beginning, most of them have four wings, whereas a bird has two, and while the wingbeats of birds range from one to fifty per second, those of insects vary from eight to an incredible one thousand per second. Hardly surprising, then, that I was curious to know what an insect is really doing during its seemingly frenzied activity of flight.

Until now, practically all research and photography on insect flight has been carried out under artificial conditions, with tethered insects suspended in a string or wire harness. This method obviously interferes with normal flight and induces unnatural air currents and stresses on the insect, rendering its performance abnormal.

Much has been learned from such experiments, of course, but many of the findings must be inconclusive, for the insects are not in natural and free flight. In my photography the only unnatural condition is that the insects fly in a studio rather than in the open sky, but always with a background of their usual foliage, and always in complete freedom.

I have taken as many as 900 exposures of a single species to achieve the result I want, for insects can be as stubborn, or as cooperative, as any sitter, but neither they nor their performance are ever controlled. Moreover, following the rigors of their photographic stint, they are returned alive and well to their natural habitats, there, no doubt, to regale companions with wondrous tales of unimagined worlds.

Insects, apart from certain hovering species, such as hawk moths and hover-flies, have not been photographed successfully in free flight, because certain fundamental technical difficulties have, until now, remained unsolved.

The first problem was the smallness of insects. This requires a much higher magnification on the negative to achieve a reasonable image size, thus imposing severe limitations regarding depth of field. In macrophotography, depth of field decreases rapidly as magnification is increased, and so the smaller the insect, the more difficult it is to get it in sharp focus from front to rear. This is aggravated in that greater depth of field is required for an insect flying than for an immobile one; the mere fact that its wings are outstretched demands more to bring into focus.

This means that in photographing flying insects I have absolutely no margin for error and am always working at the extreme limits. If the insect is not located exactly in a predetermined plane, then one extremity or the other—the end of the antennae or the tip of its abdomen—will be out of focus.

The second problem was that insects move at relatively high speeds for their size, along quite unpredictable flight paths. How often do you see an insect flying in a straight line? Most insects zigzag indiscriminately, without apparent rhyme or reason. I had to develop an optical-electronic system for accurate detection of the insect in a precise plane of space. And a further system for opening the camera shutter at that exact moment of detection.

A normal camera shutter takes at least 1/20 second to open, so that an insect flying at five meters per second will have moved ten inches past the point of focus during the time it takes the shutter to open—only good enough for a blurred insect at best, or more likely nothing at all in the picture. So shutter delay had to be kept to an absolute minimum.

This was achieved by designing a special rapid-opening shutter which could be clamped in front of the lens and used instead of the focal plane shutter of my 35mm Leicaflex camera, which was left permanently open. The special shutter had to be capable of opening and firing the flash within 1/500 second, putting it beyond the range of any practical commercial product. The success of the shutter was, of course, imperative to the whole project.

A third and more obvious problem was the need to eliminate all image blur due to both the forward motion of the insect and its wing vibration. The only way to arrest this movement was to use a very high-speed flash. Conventional electronic flash was far too slow, for with most insects it is necessary to reduce the flash duration to at least 1/20,000 second, where the normal conventional unit has a duration of between 1/400 and 1/1000 second. So the flash unit, too, had to be an original, as even the specialist 1/5000 second units used by some bird photographers would be quite inadequate.

So, if I was ever to take really good photographs of insects flying, I needed to design radically new equipment to overcome these problems. It now dawned on me that the only camera part I already possessed, suitable for the job, was my lens—a somewhat minimal beginning.

Inquiries in England, on the Continent, and in the United States confirmed that there were no commercial electronic shutters or flash units that met my requirements. Nobody made suitable rapid-opening shutters, and though a few companies manufactured flash units for specialized scientific purposes, none was at all suited to insect photography. One produced flash durations down to 1/1,000,000 second, while others were designed for a succession of flashes, but none was at all powerful enough, and they all suffered from various other disadvantages.

I needed a combination of two conflicting properties—speed and power. The so-called computer flash units on the market, although capable of high speeds, do not deliver nearly enough power at those speeds.

One London electronics company offered to do some experimental work, against an advance of four hundred pounds, with no great hope of success, and this brought home to me that not only did I need unobtainable equipment; I urgently needed money for the development of the project. Teaching photography and living from the proceeds of free-lance work hardly made up the financial base for the specialist electronic know-how I was seeking.

At this point of high hope and low finance, sometime in 1969, the beginning of real progress came from the heart of the photographic industry. Each year Kodak offered awards in the form of financial grants for valid photographic projects, and I applied for one. I found the section of the application form which required a detailed cost estimate for the project somewhat difficult to fill in, but I fudged it as best I could and a few months later was delighted to be named recipient of a Kodak Award, in the sum of four hundred pounds!

Being thrifty by nature, I did not let the coincidental amount of the award return me hurriedly to the electronics company. Instead, I took it as a good omen of just reward where reward was due, and decided to look for additional assistance.

My first visit was to the National Physics Laboratory at Teddington, where I spent a day with a number of scientists, engineers, and scientific photographers. They were interested in the problems, and made some useful comments, but nothing came up that promised any real help. As I was leaving, one of them suggested I get in touch with the Central Unit of Scientific Photography at the Royal Aircraft Establishment, Farnborough. "They might just come up with something down there."

My visit to the Royal Aircraft Establishment proved to be a somewhat cloak-and-dagger affair. Special security clearance well in advance was necessary. In due course this was forthcoming, and early one spring morning I piloted a light plane through miserable visibility to a radar-assisted landing at an unfamiliar airfield. I felt such an arrival was an appropriately stylish way to appear for my interview, but it did not daunt the security police, who painstakingly searched my briefcase and made me complete endless forms before confirming my clearance.

Eventually I was escorted to the Central Unit of Scientific Photography and to a meeting with Ronald Perkins, an electronics scientist also much interested in photography. I quickly discovered that Ron Perkins is a man of few words, an excellent listener, and a nonstop thinker.

I rambled on about my hopes and ideas and work to date, and he offered that he himself had thought of developing a high-speed flash for his own bird photography. Soon we were on common ground, and I realized I was talking to a man sympathetic to everything I was trying to achieve.

When I finished, Ron said that the time and money needed for the development of suitable apparatus through C.U.S.P. would be prohibitive, but that he might give me a hand himself in his spare time, though he would first have to check it out at home. After lunch with his wife, he returned. "It's OK for me to help you," he said quite simply. It was a wonderful offer, and I flew back scarcely able to believe my luck.

One week later I went over to Ron's home and we began discussing basic design specifications, size, weight, possibility of field use, maximum safe voltages, mains versus battery operation, number of flash heads, reflector design. I had already worked out my speed and power requirements, based on my stipulation that Kodachrome II film (ASA 25) was to be used for the project. This was essential for the definition and color fidelity I needed. Unfortunately, as this is about the slowest color film on the market, a great deal of light output is required if the film is to be exposed sufficiently at the small apertures necessary to obtain adequate depth of field. This power problem is aggravated by the fact that the higher the image magnification, the more light one requires.

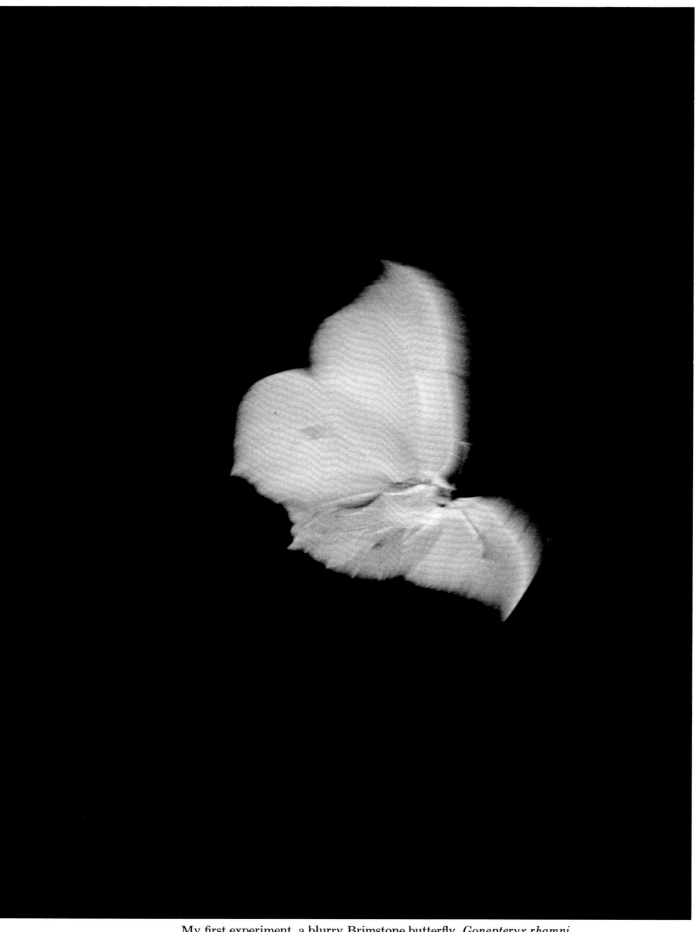

My first experiment, a blurry Brimstone butterfly, *Gonepteryx rhamni,*
taken with conventional flash at 1/1000 second. See also plate 21.

Schematic diagram

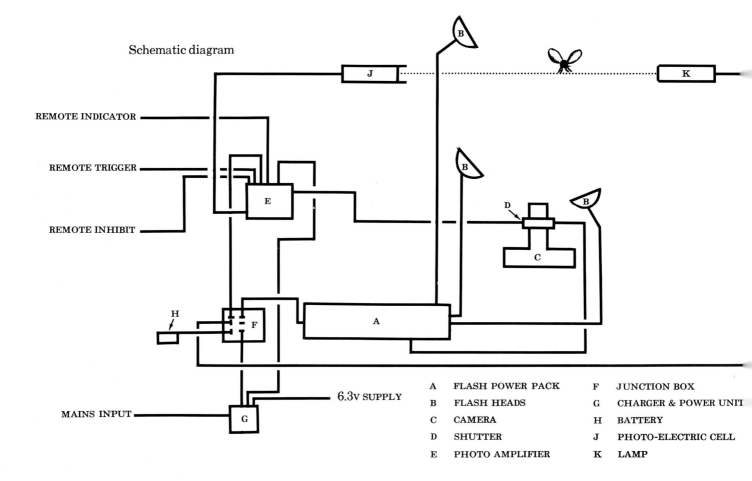

REMOTE INDICATOR

REMOTE TRIGGER

REMOTE INHIBIT

H

F

E

J

B

B

D

C

B

A

MAINS INPUT

G

6.3v SUPPLY

A	FLASH POWER PACK	F	JUNCTION BOX
B	FLASH HEADS	G	CHARGER & POWER UNIT
C	CAMERA	H	BATTERY
D	SHUTTER	J	PHOTO-ELECTRIC CELL
E	PHOTO AMPLIFIER	K	LAMP

Ron Perkins

The basic studio equipment at an early stage, showing flight tunnel and an untidy array of apparatus which has since been much streamlined.

Desert Locust, *Schistocerca gregaria*. See also Plate 9.

Common houseflies, *Musca domestica*, mating in flight.

Common Housefly, *Musca domestica*.
Separate photographs showing various stages in take-off. See also Plate 69.

Common Housefly, *Musca domestica*, taken with multi-flash and showing touch-down technique on ceiling.

Common Wasp, *Vespa vulgaris*. See also Plates 61, 62.

Green Lestes Damselfly, *Lestes sponsa*. See also Plates 5, 6, 7.

Common Sympetrum Dragonfly, *Sympetrum striolatum*. See also Plate 2.

For my purpose, duration of flash would have to be no slower than 1/25,000 second, and the light produced by one flash head 20 inches away from the subject sufficient to expose Kodachrome II film at f/16 at an image reproduction ratio of 1:1. In addition, the unit would have to be capable of operating three flash heads. Lastly, the equipment should be portable, so that I could take it from one location to another. Raising one eyebrow barely perceptibly, Ron said that these requirements were enough for him to be going on with!

With all electronic flash, including computer flash, power and speed are in direct opposition. It is relatively easy to produce a short-duration flash with a low-power output, but not power and speed simultaneously. The characteristics of electronic flash revolve around the capacitors which store the electrical energy, and the flash tubes through which this energy is discharged. The duration of flash, its power, the voltages required, and the size and weight of the apparatus are all keyed to these two components. The trick would be to find the right combination for our purpose.

Over the following weeks we purchased tubes and capacitors from Germany, the United States, and England, and we spent most of the next year experimenting with different combinations of these, until we finally arrived at the required balance of very bright and very fast flash, achieved with an American capacitor and a tube from Great Britain.

Ron's next task was to design a suitable power unit to operate them, while I developed a precision optical system to detect and react to the insects in flight. Basically my plan was to use a light beam, half a centimeter in diameter, directed toward a photoelectric cell, with an ultrasensitive amplifier capable of responding to the smallest insect, located approximately six inches from the light source.

Since it was most unlikely that an insect, in its erratic flight path, would actually break this very thin beam of light, I arranged a series of reflecting mirrors on each side of it, so that the beam was virtually duplicated many times over. This meant that the insect would have to fly through a barrage of reflected light beams rendered so sensitive by the amplifier that even the shadow of an antenna falling across any one of the reflections would be enough to fire the camera. (Figure 12 Page 150.)

I spent a lot of time on this design and finally showed Ron the results of my labors. I could almost feel him shudder when he inspected my amplifier, which he then replaced with one he had been quietly working on. It was about fifty times more sensitive than mine!

During the next few months we improved the detecting system to the extent that it responded to a human hair passed as fast as I could move it through the light beam, with a delay approaching 1/1,000,000 second. If a human hair "flying" through the beam triggered the camera and flash, then an insect's antenna should do the same. We were heading in the right direction.

Consistently we sought to improve the performance of all components. One refinement needed a special halogen point light source of a type only in prototype form at that time. I ordered it and had all but given up hope of ever receiving it, when, three months later, a huge articulated truck roared down my narrow country lane, knocked a sizable limb from my favorite oak tree, and came to a grinding stop. The driver jumped down and solemnly handed me a two-inch-square package. My precious lamp!

Before testing the detection system on insects, I needed to devise a way to make insects fly through the light beams to trigger the flash and shutter. It seemed logical that most insects would fly toward an exit of light, so I made a tunnel through which they could make their way to the great outdoors again, photographing themselves as they went. So impatient was I to try this out that I decided to test it without Ron's flash or the necessary shutter, neither of which were yet completed.

My first experiment was with a Brimstone butterfly, a Leicaflex SL with a 100mm Macro Elmar lens, conventional flash of 1/1000 second, and open-flash technique. This meant that when the butterfly triggered the flash as it flew through the light beams, I had to open the camera shutter manually immediately before the flash fired. I spent a whole day encouraging the debut of the Brimstone, and used up thirty-six exposures, of which only three had any image at all. But it was a start, albeit a blurred one, and the photographs I eventually obtained with this crude method of flash and shutter at least proved the efficiency of the optical light trip and the photo amplifier.

Soon after, Ron phoned to say that the electronic flash power unit was ready for testing. In a diffident, almost apologetic manner he pointed to the flash heads, an oscilloscope, the power unit, and a maze of test meters and wiring. He painstakingly explained every detail of the circuitry, quoting all the voltages and currents indicated by the meters. He added that in the wrong hands the equipment could be lethal, for at this experimental stage it was capable of delivering a current of hundreds of amps at several thousand volts.

The power involved was evident on the discharge of the flash, which exploded like a rifle shot. Later on in the week, when I was testing the apparatus at home, a neighbor came over to find out what all the noise was about, fearing a blood feud! Subsequently we modified the circuit and reduced the noise level of the flash explosion.

On test the electronics worked perfectly, but light from the flash heads was uneven, so we had to redesign the reflectors. One problem with short-duration flash is that color balance is affected by "high intensity reciprocity failure."

Tests with different filters built into the flash heads duly corrected this.

We now had flight detection and we had flash, but the next few months were the most frustrating of all, for I still could not begin serious photography until we developed the rapid-opening shutter. The project had already taken more than eighteen months.

Our first efforts to improve the shutter-opening speed to the required 1/500 second were in some contrast to the electronic wizardry employed in the solution of the flash problem. We reverted to basic engineering principles to accelerate the opening time. A system of elastic bands, not surprisingly, proved cumbersome, inelegant, and inefficient, so we tried a slightly more drastic means—an explosive device. In a sense this did work, but the cost per exposure, the severe vibration to the camera, the slow resetting time, the limited shutter life, not to mention the noise and the smoke, forced us to abandon this idea, too.

Ultimately we developed an electronic shutter with an opening time about 1/450 second, which has since proved fast enough in most circumstances, but not fast enough for some of the most rapid insects.

By spring 1971 we had a highly usable, though still unwieldy, array of equipment, with which I began photographing insects, while Ron continued to refine and streamline its components to meet his ever-increasing standards.

By far the most important refinement was the development of the multiflash. This device made it possible to fire a sequence of three flashes with a predetermined time delay between each, so that a series of three images in the same film frame could be recorded. The technique not only shows different wing positions within a relatively short space of time but also makes possible the study of insect take-off and landing behavior, never before seen.

During the summers of 1971 and 1972 I photographed more than 100 species of insects in flight, using more than 15,000 exposures. In the autumn of 1973 I flew to Florida, thanks to the generous help of the Reader's Digest Press, with a superbly portable box of tricks, to photograph insects in Everglades National Park.

With the help of the Park authorities I moved into a motel at Flamingo Bay, the southernmost tip of the Everglades, and assembled the apparatus, much to the dismay of the maid, whose customary daily routine had to be drastically revised. That anybody should travel three thousand miles to photograph the pests of the swamps seemed to her an illogical move at best.

For a naturalist photographer, Everglades National Park is a most exciting hunting ground, and the temptation to photograph some of the many species of magnificent and rare birds, as well as the insects, was more than I could resist. However, the fascination of getting to know and photograph new insects soon preoccupied me, and I spent the next six weeks crawling happily through the hammocks and subtropical vegetation, searching for the species I wanted.

The Park rangers, after first displaying some suspicion about my activities, were soon directing me to the most promising areas. During my stay I took nearly a thousand shots, going in search of insects first thing in the morning and during the softening light of early evening, spending the hottest time of the day photographing them in air-conditioned comfort.

One of my most vivid yet at the same time most frustrating recollections is of my room filled with Zebra butterflies dancing lyrically around the ceiling but refusing to cooperate in any way. They entranced me for hours before I was finally able to capture the lovely, graceful movement of their flight.

The equipment worked perfectly, after an initial scare that had me extremely worried. When I first assembled all parts of the system, I connected the power supply and nothing at all happened. A nerve-racking few hours were spent inspecting transformers, rectifiers, and poring over circuit diagrams until I discovered the one mistake Ron had made in the entire project. He had connected two wires in the power supply the wrong way around! With a sense of great relief, I switched them over, and all was well. I had no more trouble with the system, or the maid, the whole time I was in the Everglades, and the equipment traveled safely both ways.

For a photographer, the whole experience was totally engrossing; for the naturalist in me, fascinating and exciting; and for the insects I photographed, hopefully a welcome change from their daily round.

What next? In the bleakness of an English February day, I believe a modest expedition to seek out the butterflies of the Upper Amazon would be agreeable. My equipment is packed and ready to go. And so am I.